This book is a series of essays by people from all walks of life and all parts of the country. It focuses on their successes with survival, service, leadership and preparedness. It is intended to put a human face on one of the nation's worst national disasters and to emphasize the power of the human spirit to inspire, motivate and accomplish. It can also be used in educational settings for history, social studies, science and service learning instruction. A companion website with photographs, updates and lesson plans can be found at *www.TheInnerLink.com.*

The authors are sensitive to the massive losses suffered by the people in the Gulf region. They have created this book as part of a continuing national healing and recovery process. A significant portion of the profits from this book will go to the authors' favorite charities and programs, including some in the City of New Orleans and the Mississippi school districts of Long Beach and Pass Christian.

Lessons Learned: Successes Achieved

Lessons Learned: Successes Achieved

✦

Be Prepared for Disaster: Advice from Katrina Survivors

Robert Gillio, MD and
Evangeline Franklin, MD, MPH

iUniverse, Inc.
New York Lincoln Shanghai

Lessons Learned: Successes Achieved
Be Prepared for Disaster: Advice from Katrina Survivors

iUniverse books may be ordered through booksellers or by contacting:

iUniverse
2021 Pine Lake Road, Suite 100
Lincoln, NE 68512
www.iuniverse.com
1-800-Authors (1-800-288-4677)

www.theinnerlink.com

ISBN-13: 978-0-595-41756-8 (pbk)
ISBN-13: 978-0-595-86097-5 (ebk)
ISBN-10: 0-595-41756-6 (pbk)
ISBN-10: 0-595-86097-4 (ebk)

Printed in the United States of America

This book is dedicated to all those affected by Hurricanes Katrina and Rita. Your courage and stamina as victims, volunteers and professionals have inspired a nation. Your actions will help train and prepare the nation for emergencies, so that others will survive, and then thrive, in the face of adversity.

Contents

Introduction xiii

CHAPTER 1 Maggie Landry 1

CHAPTER 2 Robert Gillio 7

CHAPTER 3 Evangeline Franklin 18

CHAPTER 4 Emily Burris 22

CHAPTER 5 Rod Schoening 26

CHAPTER 6 Garrett Book 31

CHAPTER 7 Tammy Lee 36

CHAPTER 8 David Ressler 43

CHAPTER 9 Greg Vogt 50

CHAPTER 10 Greg Porter 55

CHAPTER 11 The Gillio Family 58

CHAPTER 12 Robbin Bertucci 74

CHAPTER 13 Ayanna Buckner 76

CHAPTER 14 Kevin Stephens 82

CHAPTER 15 John Tardibuono 93

CHAPTER 16 Kirk Sharp 97

CHAPTER 17 King Lam 104

CHAPTER 18 Vivian Ward 108

CHAPTER 19 Chris Porter 110
CHAPTER 20 Nancy Burris Perret 114
CHAPTER 21 Evangeline Franklin 122
CHAPTER 22 Summary and Recommendations. 134
About the Editors 143

When Hurricane Katrina struck the Gulf Coast on August 29, 2005, the world, as hundreds of thousands of Americans knew it, came to an end. Two doctors in different parts of the country became involved in the response and recovery efforts, and through their eyes—and through the experiences of those they met along the way—we learn that heroes live among us, and that many successes were achieved by courageous and caring people with the knowledge and skills to make a difference.

Good things happen when people step up and work hard. The people and stories in this book reflect the best of America. We celebrate their successes achieved in the face of adversity.

This book is not a conventional lessons learned emergency management text, which critiques what went wrong. While those hard lessons will need to be discussed—and changes made—we instead now seek to focus on those things that went *right*, and to celebrate the successes achieved by ordinary people in an extraordinary time. This book is about one storm in one region, but it is hurricane season on both coasts of America for 6 months a year. It is also accident and terrorist season everyday, everywhere. When a disaster happens, it does not discriminate in who it affects and what it destroys. It is a great equalizer of rich and poor in the wind and wet, injured and scared. Some may be better positioned to survive but all are at risk. It is the authors goals that the celebration of the successes and lessons learned will motivate the reader to pay special attention to the last chapter and work to become part of the solution and get trained and prepared to respond to their own needs and those of their neighbors in an emergency.

Then, we can all take what we've learned to plan for, respond to, and recover from adversity in our own homes and communities.

We all need to be prepared

Introduction

✦

Lessons Learned: Successes Achieved

When terrorists attacked America on September 11, 2001, the country was in shock. A sense of security was lost forever. We had been attacked on our soil. People had died. We saw it unfold, live, on television. Our country's airplanes crashed into buildings we had worked in, visited, or seen. We watched helplessly as these businesses and buildings caught fire and then collapsed.

In the days and months that followed, the word "homeland" came into common use. The country focused on the victims and the stories of heroism. An overwhelming outpouring of patriotic feelings and symbolism swept the land, accompanied by acts of generosity and volunteerism. Everyone wanted to help.

I was lucky enough to be part of the response and recovery team in the early stages of the disaster. As a lung doctor, I was able to help New York Police Department workers who were having chest tightness and possible heart disease issues. I found myself eventually leading a care and research activity, using improvised technology that I had designed to teach science students about biology and anatomy. When all was said and done, the team had taken care of 1,800 police officers, and had arranged for follow-up care and an ongoing study of those officers' health.

These were some of the most gratifying moments in my professional career. They solidified my decision to reduce my clinical practice so that I could focus on public health issues in schools and communities. I wrote a book called *Lessons Learned at Ground Zero*, and it led to my involvement with the White House Office of Homeland Security. I found that, whereas the 9/11 attacks initially left my family and I feeling weakened, we felt empowered when we became an active, helpful part of the responder/volunteer community. We believed that we were doing the right thing, and that we were part of history. I began to build solutions to help respond to the next major disaster.

That major disaster was Hurricane Katrina.

This was a storm of historic proportions. It was a Category Five at landfall, with wind speeds of eighty to one hundred forty miles per hour, storm surges of twenty to thirty feet high over the length of more than one hundred miles, and rainfall in excess of fifteen inches, in some places. As with 9/11, most of us saw the events develop on television. Many of us felt the familiar helplessness. As a nation, we raised money, we wrote checks, we sent supplies, but the disaster worsened. Things went wrong—very wrong. A dike burst. People were trapped, and we watched in our living rooms as citizens begged for help—and then died. We saw a city flooded.

For the most part, we did not see the adjoining catastrophe—that is, the tsunami-like destruction along the entire coast of Mississippi. Everything was wiped out for ninety miles, including houses, cars, schools, towns and people's lives. Then, another hurricane came, bringing a major urban flood. The agencies that respond to such crises were overwhelmed. They had never faced a major disaster that spanned such a massive geographical area.

We learned a very important lesson from this tragic experience: Every emergency is, essentially, a local emergency, and the local community or individual must be prepared to respond effectively.

Some people have spent valuable time trying to assign blame, publishing finger-pointing media accounts and technical post-action reports. That is not the purpose of this book. This book will instead focus on what went *right*. It is an important task because, in reviewing what *did* work well in response to Katrina, we can learn how to modify planning for the next event. We can also celebrate the ingenuity, heroism, intelligence, and citizenship of the people who courageously responded to this disaster.

Perhaps the greatest gift I have received in my professional life is the pleasure of meeting or hearing about the people who have contributed essays to this book. During the crisis, they became the de facto first responders, saving themselves and then scavenging resources to help others. They were part of what I call the

Freedom Generation, a group of people who are free of ignorance and can independently care for themselves and others. They are free of fear of the unknown, because they make good decisions, and they know how to react to situations that they cannot control. They are my role models, my heroes, and the people to whom I turn in order to learn how to properly train our citizens to acquire the health, safety, and emergency response skills they need in the event of a disaster.

A disaster strikes indiscriminately, regardless of age, socioeconomic status or skin color. The stories included in this book did not make that distinction, either. They were written by persons of various ethnic and cultural affiliations, and from all walks of life: Young and old; rich and poor; those who hold powerful positions in society, and those who are struggling just to survive. Each writer has something to teach us. Each has faced challenges with courage, leadership and a sense of hope—a cross-boundary approach that can be used to overcome any adversity.

After reading their stories—and viewing some of their pictures and video interviews online at *www.TheInnerLink.com*—I invite you to read the book's closing remarks and practical guide. It is our hope that you will then be inspired to become a member of the Freedom Generation, leading yourself, your family and your community into "Successes Achieved".

1

Maggie Landry

Maggie was a Louisiana State University graduate student when Katrina struck in 2005. She was also a homeowner, wife and mother who faced many struggles brought on by the storm. As you read Maggie's essay, you should note the things that she did to make the area's difficult recovery phase much easier. Following the last chapter, we will discuss these actions, and attempt to put them into a practical plan for you for your family.

—Dr. Robert "Dr. Rob" Gillio

My husband Kai and I moved to Biloxi in August 2003. Both of us were from southern Louisiana, and, after five years in the desert Southwest, the coast, and the oak trees and azaleas were comforting to us. It was like coming home. Eight months after we arrived, we bought a beautiful Victorian home just one block from the beach. Kai was working and I stayed home with the girls, driving to Baton Rouge twice a week to attend classes at Louisiana State University. I had started work on my Ph.D. in January 2004. We were living on the coast; we had a beautiful home; the girls had lots of new friends, and so did we. We knew we were right where we wanted to be.

Hurricane Katrina was already making news when I took our two daughters to my mother's house in Baton Rouge to see the children's stage show, *Disney Live! Winnie the Pooh*. By the time the curtain went up on Friday, August 26, I had already made plans to drive back to Biloxi alone the next day so that Kai and I could board up the house. The girls, then four and two, were oblivious to our concerns; they simply looked forward to being at Grandmother's house without Mom around. All day Saturday and all day Sunday, we covered our home's windows with plywood. We picked up the lawn furniture, toys, bicycles—and anything else in the yard that had the potential to become a projectile in hurricane force winds—and put them in the shed or in the house. Around 9:30 on Sunday morning, I remember looking up at the blue sky and seeing white clouds moving in that familiar circular pattern you usually see as a satellite image on television. Only this time, I was under the pattern; I knew that I was seeing the extreme outskirts of Hurricane Katrina as she headed our way.

We had already gone through two hurricane preparations since we'd moved to Biloxi. It is an odd experience. You know that the storm is coming, and you know about when it will hit, but you can never be sure of exactly *where* it will hit, or in what direction it will go from there. As I had done twice before, I got out a large box and put everything that could be defined as "irreplaceable" in it: Our wedding pictures, the baby books, photographs of my parents from their wedding, and photographs of their parents' weddings. All the while, I was saying to myself, "The house will be fine. It's not going to be that bad. I really don't need to be doing this." But I did it anyway, because it always makes me feel better—just like it makes me feel better to board up the windows, clean up the yard, and turn off the gas. You hope that what you are doing is unnecessary, but you do it, just in case of the worst.

The rain started late in the afternoon. We had done all we could do. Because all of the windows were boarded up, the house was dark as we showered, ate and got ready to leave. I packed up some clothes, some toys for the girls, their immunization records—just some things that I thought would be useful for a few days' stay in Baton Rouge. I gave my neighbors keys to our house, since they had decided to ride it out. If the water rose, they could go to our upper floor, and stay dry.

Lots of people in our neighborhood were staying. Some of them had been through Camille in 1969, and had seen her flood homes along the beach. We were all further inland, and on higher ground. How bad could it be?

We left Biloxi around 9:00 PM Sunday, August 28. By this time, the nightmare of evacuation traffic was over and we drove the 150 miles easily. As we drove, we listened to radio personality Spud McConnell take calls from people all over the Gulf Coast who were, at this point, just waiting for the dawn.

I woke up early on August 29 and turned on the television news. The Mississippi Gulf Coast was getting the worst of Katrina. Eventually, power went out in

Baton Rouge. We listened to a battery-powered radio for updates, but no information was coming in. All we could do was wait.

At about 2:00 AM on Tuesday, August 30, Kai and I heard on the radio that Route I-12 was cleared. They also announced that Mississippi was "closed," and that no one would be allowed in. All patience lost, Kai got into the car and left to see what we had left. I hated to see him go, but I also badly wanted to know what had happened in Biloxi.

He was able to call my mother's house, using a satellite phone he had borrowed from someone who worked at MSNBC. He said that we still had a house, but that Biloxi was in shambles. On Thursday, September 1, he and I made anther trip, and I got to see it for myself. In two short days, roads had been cleared and things looked better—but "better" is a relative term. Our house was OK; there was damage, but there would be plenty of time to assess it later. Everything that had once been stored or lost under the house had blown into the backyard. Tree branches were everywhere. It was the same for everyone immediately around us. Neighbors told us the Gulf rose all the way to our front fence and then went back, as if it were an ordinary tide.

I started to walk south. The roof had blown off one house and water came into the next several houses. The two homes behind the beachfront houses were crumbled, and the beachfront houses just weren't there anymore. Overall, Beach Boulevard was bare, and so were the trees. Instead of leaves, they were covered with people's clothes and personal property. Just four days before, the area was filled with shady oaks and gorgeous homes. Suddenly, you could see for miles.

We spent a good portion of the day there, cleaning up around the house and yard. It seemed like no matter how much we did there was still so much to be done.

We drove around downtown Biloxi that afternoon, and I simply cannot describe what it felt like to see what I saw. The streets I knew so well were now completely and totally destroyed. Buildings were collapsed, rooftops were in the streets, and the interiors of homes and businesses were strewn all over the place. The casinos had relocated, and the bridge to Ocean Springs had collapsed. I just couldn't make sense of it all. Residents were wandering around, stone-faced, and the media was frantically setting up equipment.

As cloudy as my mind was by this point, I realized that we were some of the luckiest people in Biloxi. Homes north of us were flooded and damaged more than ours. Peoples' businesses were wiped out; their way of making a living was gone.

When I went back to Baton Rouge that day, I didn't know what to think. Kai stuck around for another day or so to clean up the yard and neighborhood. Without lights, air conditioning or any running water, he only stayed one night and then came back. Baton Rouge had become chaotic. Lines at the gas stations seemed to go on for miles. The stores that were open had limited hours, and their shelves were bare. Nothing was normal. Thousands of New Orleanians had come to Baton Rouge facing their own uncertainty and lots of unknowns. At least I had been able to get to my house and could see what had happened there; they would not be allowed back into their city for months.

Kai lost his job, as had nearly everyone along the coast. Not sure what his employers had planned for the future, we decided he would go back out West to find a temporary job, since he still had a lot of contacts out there and friends to stay with. The girls and I would stay with Mom, and I would remain enrolled at LSU for the semester.

I was able to secure a spot for our oldest daughter at the LSU School of Human Ecology Preschool. Thinking about it now, it seems like things fell into place, but at the time, I had no idea if I was doing the right thing. We just we made our decisions, carried them out, and hoped for the best.

The day I unpacked my clothes at my mom's house can only be described as weird. I was thirty-three years old, moving back into the house I grew up in, with my own two girls. My dad had died nineteen months before Katrina hit, and I was not used to his absence. I was in a completely familiar place, and yet I was completely lost.

Our daughter came down with pneumonia right before her temporary school started. She was so sick, and she did not want me to leave her side. My mind was racing with thoughts of our unknown future, and there she was, lying on the couch. She was weak, coughing and miserable. Our two-year-old demanded my attention at the same time. It was hard for me to focus on them, but tending to their needs helped distract me from thoughts of how miserable I was.

In a few days, our daughter recovered and went to school. I think the familiar routine was good for her. She was very aware that this was not her old school, and I am not sure she knew what to make of it. She was quiet and reserved—not the way I knew her to be—but then again, she was not alone. All of her classmates were experiencing Katrina, somehow. Some were from New Orleans. Many of those from Baton Rouge had relatives from New Orleans living with them. The children talked about the storm, acted out the storm and coped with it in their own ways. Thanks to the staff there, our daughter eventually came back into herself. It was a thrill to watch. I learned that, as her mother, I can't be everything to her, and I shouldn't be. The school staff gave her an experience that I could not give her, with everything else going on in our lives.

In school that semester, I had to read, of all things, Kuhn's *Theory of Scientific Revolution*. This is a difficult topic in the best of times, and practically impossible to grasp in the worst of times. My first class of the semester was held ten days after Katrina hit, and I burst into tears when asked if I had anything to say about Kuhn. I was too exhausted to even be embarrassed. My brain could not process any more information. I had no idea why I was even sitting in that class. Emotionally, I was a wreck. Kai had lost his job, and we had no income. Our family was separated indefinitely for the next few months. Our house needed a new roof and countless other repairs. If it were not for my family of origin, I would not have been able to feed my children. All I saw when I closed my eyes at night was devastation and destruction and debris. So much of what was part of our lives had either been severely damaged or completely destroyed. The library, the post office, restaurants we knew, the pediatrician's office, the parks, our own neighborhood—all had been reduced to backdrops for journalists' sappy stories about a town they had never before visited.

The year since then has been a whirlwind, and my professional life has changed dramatically. I completed my semester at LSU. In December, Kai returned, and all four of us went home. Both girls went back to their old school in Ocean Springs, Mississippi, and I got a job at the Air Force Base. Before Katrina, when I had been a student and a stay-home mom, it was hard. Unfortunately, it took this disaster for me to take time off from school and give myself a break. The job was a good career move. Kai is now a full-time stay-home dad. I miss being the one who gets the girls ready in the morning and stays with them all day, but life takes sacrifices. I am far more aware of that fact, now.

On a personal level, I am better able to appreciate what I have. I consider myself extremely lucky when I think that, with a slight turn toward the east, Katrina could have been far worse on us. I don't have to walk far past our home to see what could have been. Even so, I still love a good thunderstorm and hurricane season does not rattle me. The next time something enters the Gulf, of course I'll take note and watch it closely. But I can't let it consume me.

I will never stay for a storm that severe. It is not worth it. What's going to happen is going to happen. It is better to prepare and then leave. The rest is out of our hands.

We were successful in dealing with this storm. We boarded up and prepared the house as best we could. And then we left. We got out of harm's way and we stayed together as a family. We had proof of health and homeowner's insurance, immunization records, birth and marriage certificates and social security cards with us. We had our box of irreplaceables just in case we lost the house.

Good things do come out of bad situations. I have learned that. I always knew I was too proud to beg, but I have learned that in the darkest times, pride has to be set aside. You have to take care of your family. I have learned that you have to let yourself and the people you love to cope with situations in whatever ways are best for each person. When you make a decision—even if you can't think straight when making it—you must trust that you are making the best decision you can at the time. Things have a way of working themselves out.

Getting through this disaster would not have been possible without the support of my family. For that, I am forever grateful. I am also grateful for the people who have come to Mississippi to give their time, energy and talents to clean up our coast and provide food and water for those who had none. I'm also grateful to those who are now helping us as we rebuild our community.

As for the people of Mississippi, I am proud to be counted among you. The Gulf Coast is a diamond buried underground now, but it will shine again.

2

Robert Gillio

I met "Dr. Rob"—as he is known in his education programs—in February 2005, during a post-Katrina medical event called Health Recovery Week. A pulmonologist who had treated New York City rescue workers just after September 11, 2001, he was there as an advisor to us.

He is a geeky doctor. He knows how technology can be used to help organize, document and train people, and he has a special interest in electronic medical records. During our five days together, he observed, advised, and then rolled up his sleeves to pitch in with the five thousand patients who needed treatment.

As you read his essay, I hope you will be motivated to think of what you can do to prepare for future disasters, and to help in your own community.

—Dr. Evangeline "Dr. Vangy" Franklin

Before the Storm

In the year before Katrina, I settled into work at my company, InnerLink, where I had assembled a team of people from all disciplines who could create health and safety and emergency response solutions for individuals, families, schools, and communities. My five daughters ranged from pre-teen to young adult. When she wasn't running the house and helping the girls grow up to be good kids, my wife, Beth, helped at InnerLink with our Project Nutrition program.

Although still a physician, I was not seeing patients frequently, because I had turned my attention to the public aspects of health. I was obsessed with finding ways to make it "cool" to make healthy decisions. I was also involved in a study of the New York City police officers I had treated after 9/11, at Ground Zero.

The company was poised to get its programs out to the nation. It was a critical time. I was asked to join a group from Mayo Clinic to help with the tsunami in Asia. I resisted, with much guilt, because I was concerned for my safety—and for my livelihood, if my company lost the momentum it was finally building.

The Storm Hits

As Katrina approached, it got minor coverage on the news. Sure, it looked big, but the network weather professionals have given false alarms for so long that we often tend to ignore the hype. Then the flood struck New Orleans and the media sent pictures of people in distress, begging for help from areas that rescuers could not seem to reach. It was alarming. Soon, we saw images of rooftop rescues, failed rescue attempts and masses of people, scared for their lives, because of the dangerous conditions in the shelters. Most of us experienced that same feeling of impotence that overcame the nation after 9/11.

When my youngest daughter said, "Dad, let's get the canoe and go down there and rescue that family," I knew that the nation was feeling the stress of this trauma very personally. Many people generously offered their help, but as a person trained in emergency response, I knew that spontaneous volunteers often land in danger themselves, adding a burden to the already overwhelmed rescuers and healthcare systems.

After the Storm

In my community, we had created a volunteer branch of the U.S. Surgeon General's office called a local Medical Reserves Corps (MRC). This was one of dozens of such organizations around the country that were set up to deal with biological, chemical or nuclear attack at the local level.

My company had proposed ways to use computers to collaboratively allow these MRC teams to have uniform communications, training, credentialing, telemedicine capabilities and resource allocation. These proposals were refused with the argument that all disasters are local, and that solutions are to be made on a local basis.

I saw the waste inherent in a system in which a multitude of organizations come up with their own solutions, reinventing different versions of the wheel that won't work together on the same vehicle. Compounding this error, I saw local communities taking the Homeland Security money after 9/11 and strengthening the silos of ownership over different aspects of disaster management, instead of doing what was recommended by my company and others after 9/11. Information should be shared on a need-to-know basis, so that integrated planning can prevent, prepare, respond and recover from emergencies. Basically, in this instance, everyone had his or her own ideas about how best to spend funds, and who was in charge. The right hand didn't know what the left hand was doing, and nobody knew where to find all of the resources needed to get the job done.

As the situation worsened in New Orleans, my daughter Maria, a sixth grader, could see that folks could have been evacuated by the unused school buses that were now sitting in flood waters. The desperate attempts at evacuation and family reunification were heroic and pathetic at the same time. The various groups and agencies within local governments, adjoining communities and the state and federal government agencies tried hard, but there was a significant lack of planning and communication that led to failure.

Books will be written about this for years to come. The Federal Emergency Management Agency (FEMA), the insurance companies and the Army Corps of Engineers will weigh in. Court and congressional hearings will explore how relief organizations were overwhelmed, or how poorly they handled huge aspects of this disaster. Read about this elsewhere. We need to move on to observations of lessons learned and success achieved.

Citizens were angry, and with cause, but their anger was misdirected and, in part, focused on the agencies they were dependent on. Did they realize that they had a responsibility to make decisions that would help optimize their own health and safety? There were things they could control and things they could not, but were they prepared to deal with both? Based on my observations in New Orleans, the answer, for the most part, was no.

One might think that folks living between two levees would have listened to FEMA's advice to put a plan for evacuation and/or shelter in place; to create stores of water and food; and to have a go-kit for quick evacuation. These kits would be stocked with a few essentials, including pertinent medical information and copies of other important papers. Did citizens take the two minutes needed to plan a unification strategy in case of separation? Did they pick a meeting place outside the region, or even set up a place to call or get together online? Were they previously trained as volunteers with basic first aid or CPR? Did they expect that everything would be taken care of by others? Although the Katrina tragedy was viewed and critiqued nationally, it was, initially, a local emergency; national aid was unavailable until local logistical support was in place. The local citizens were not well served by their own actions.

I do not mean to deny the unfathomable suffering that was beyond anyone's control. However, in many cases, the pain and suffering could have been mitigated and dramatically reduced with some simple planning. Leaders in the community and many citizens had access to information needed to make those plans. There would have been less pain and suffering if the community had followed the advice of experts who had analyzed responses to previous disasters.

The leadership shown in Harrison County, Mississippi offers an example of difficult, but good, decisions made in a time of crisis. County Supervisor Marlin Ladner was with us when we saw the destruction for the first time. As he explained the situation, he was asked about the evacuation order. He confided that it was a hard decision to make, explaining that if officials cry wolf too many times, the public won't listen to the warning. As Katrina approached, however, he ordered evacuation, and most citizens listened. As a result, relatively few deaths (238) occurred along the ninety miles of coastline, compared to the greater number of losses suffered in New Orleans. The leadership and citizens achieved a great success with this action.

The lesson learned is that we all need to take personal responsibility to plan for possible crises so that we are able to take care of ourselves in the event of a disaster, and to make sure that we are not a burden on an overwrought or ineffective system when disaster strikes. Then, we need to pitch in to help others.

Professional Successes Achieved

The Medical Reserves Corps leadership realized that Katrina was like no other emergency that it had seen or anticipated. In terms of geographic scale, the Ground Zero site in New York was a mere mosquito bite in comparison. In New York, the local hospitals were a bit chaotic for a few days, but they didn't close, they did not have damage, and, in fact, they ended up with increased incomes and huge grants in the aftermath. In New Orleans, Charity Hospital and its few outpatient clinics ceased to exist or to function. Half the doctors and other healthcare workers had evacuated out of town, and their clinics and offices were destroyed. There were victims who needed primary medical care due to heart failure, emphysema, childbirth, diabetes and cancer—primary care that was instantly gone. Then, of course, there was the trauma of broken bones and thousands of lacerations caused by working in the flood debris.

An email asked the local MRCs to send volunteer physicians and other health-care providers. The instructions for deployment were rather sterile and administrative, and they overlooked the lesson we learned in New York City, which told us that volunteer rescuers often get sick, are injured, have heart attacks, and need help, too. Because they did not have health information with them, and because the medical infrastructure on the Gulf was severely damaged, it would be difficult to help them when they needed it.

I placed a call to the Commander and proposed that InnerLink, through our local MRC, would offer the use of our Health Passport personal health record for any responder who wished to use it. The Surgeon General's staff agreed that it was a good idea, but that they could not pay for it—nor could they mandate its use, because the MRCs had local control and autonomy. Moreover, since this service would be an unanticipated expense, it would require a competitive bid process and several months to get a contract in place. (This was a disappointment. Although we are a small company in need of revenue, we had spent thousands of dollars over the years supporting our local MRC. We received little company business as a result, but we felt that we had done the right thing for our community.) My personal success was that I convinced my company's board of directors to donate this service, and I convinced the commander of the MRC to accept it. Small companies and individuals can make decisions quickly and act decisively. The MRC leadership took a risk, but had the courage to allow this simple tool to be used by the doctors it deployed.

Subsequently, hundreds of MRC volunteers self-registered and then went into harm's way, carrying with them important personal health information in a secure online/on-person format. After the urgent times following Katrina, I received a hand-written note from Richard Carmona, the Surgeon General himself, thanking us for this act of patriotism and leadership.

Once a person has experienced a large-scale disaster, he or she seems to take on an aura of disaster magnetism. By this I mean that there are people out there who want to help during a disaster and, because they know of my experience aiding others in emergency situations, many of them call me and ask how they can help.

What they may not have realized is that many of the successes I achieved following Katrina occurred *before* I made that trip.

An example of this success from home is the Lutheran Disaster Response (LDR), a faith-based group in Pennsylvania that was piloting InnerLink software to train, credential and organize its members in the event of a deployment. When the disaster occurred, InnerLink facilitated the use of software to help manage the LDR's response, and to help organize hundreds of other volunteers. For me, the emergence of faith-based organizations in disasters is the American spirit personified. The volunteers stepped in with time, talent and resources that were far more effective than the efforts put forth by our government of professionals. They were a success in many ways.

Following a discussion with one of the leaders of LDR about the upcoming deployment, I went to an art fair, where a series of paintings caught my eye. They depicted children from rural Mississippi, and they were hauntingly sad. I asked the artist, H. C. Porter, if she would consider letting me display her art in a poster, to be used to teach the nation about post-traumatic stress disorder and other emotional issues commonly caused by a widespread disaster. She was happy to help. She stated that some of the children in the paintings lived in hurricane-torn areas, and of those children, some had died. These posters were eventually created with the aid of Psychologists that focus on the psychological health of traumatized children. They are available for free and can be downloaded from the InnerLink website (*www.TheInnerLink.com*). Now they will be used for awareness in New Orleans and elsewhere, and the proceeds from sales will support psychological services for victims.

Personal Successes Achieved

A co-worker knew of a group of local students that had raised funds to tend to the emotional needs of students in Gulf Coast communities damaged by Katrina. They planned to travel to those schools, to help in person.

The leader of the group, Matt Cooper, is an energetic teacher who can take a student's question and turn it into a teachable event bigger than life. One student said to him, "I wonder what the Gulf Coast students are doing for homecoming, now that their homes and schools are flooded. Senior year is supposed to be fun and full of meaningful memories." It was a poignant observation, made by a teenager about other teenagers' disrupted, difficult lives. Realizing what an important educational experience this might be, I asked if my daughter Amy could go along with the school group. Matt said yes—if she would make a video or photo documentary.

Amy encouraged me to accompany her, and we went to Long Beach and Pass Christian, Mississippi. We saw destruction beyond belief and lived with people who were experiencing great stress in a tumultuous time. Somehow, the story of America's own tsunami in Mississippi never reached our televisions in a way that effectively depicted the scale of this biblical-scale destruction. For ninety miles along the coast there was nothing left standing.

We went there to help others feel better, but we came away feeling strengthened by the courage and determination of the area residents. As my daughter Amy observed in her documentary DVD, the experience taught us that the material things lost in the disaster are important, but not nearly as important as family, friends, and a country where strangers travel far to help "muck out" flooded homes, provide a celebration of homecoming for those who are trying to find their next home, and leave behind thousands of dollars of donations.

I am proud of the successes I achieved as a citizen of Lancaster County, Pennsylvania who helped to send volunteers and donations to the Gulf region. I am also proud of the successes achieved as a father, whose daughter had the courage to travel to a disaster zone (and the insight to force me to join her). Her documentary DVD is helping raise money to aid the recovery. She has also created an award-winning research study documenting the increased incidence of post-traumatic stress disorder in high school students in Long Beach, as well as in those who experienced the pain of the disaster remotely, via television newscasts. She has since returned two more times to the area.

She and her sister Sophia were the lead fundraisers in their Pennsylvania high school, Conestoga Valley, and the money they helped raise was used to build three Habitat for Humanity houses that were shipped to hurricane victims. Another sister, Anna, made two volunteer trips to the Gulf Coast while her college was on break. Last Easter my wife, Beth, joined Anna and Amy at Mississippi's God's Katrina Kitchen, where they all helped feed and clothe two thousand hurricane victims and volunteers. As I write this, I am awaiting the arrival of a Mississippi teacher and her daughter who wanted to personally thank some of the people "up North" who had helped them in their time of need. What a great country and volunteer spirit!

The amazing success in the aftermath of Katrina is that thousands of volunteers were empowered to understand the absolute need to prepare their families and communities for possible future emergencies. In the process of helping they, too, were helped. They also became motivated to be sure their communities weren't caught ill prepared.

I had the privilege of making several additional trips to the region, meeting people whose stories have changed me forever. Some of them have provided essays for this book. They include the following:

Leisha Pickering, wife of Mississippi Congressman Chip Pickering, oversaw a large donations effort near Jackson, Mississippi that now flows through a multi-thousand-square-foot warehouse and is staffed by a team of volunteers. Because of her work, dozens of truckloads of donated goods are able to reach the affected communities. The program has grown So large and efficient that FEMA handed part of their job over to her HANDS [Helping after Natural Disasters] Foundation.

Greg Porter knew he was able-bodied, and that if he showed up, he could help. He took a few supplies, a grill and some hamburgers to Mississippi, headed south of the tracks, and set up camp on the beach. The smell of his hamburgers attracted a few folks that first night. He never really left, and his campsite is now a complex called God's Katrina Kitchen. It is a collection of huge circus tents and donated portable buildings built by the Amish of Lancaster, Pennsylvania. Like the loaves and fishes in the Bible story, the nation's faith based community keeps supplying that grill. It now feeds two thousand victims and volunteers three meals a day, and it distributes clothes and supplies sent by thousands of churches and the HANDS Foundation.

The United States Coast Guard was especially inspirational to all of us. The Coast Guard led the rescue of at least 2,580 people in joint agency cooperation; in fact, on the Gulf Coast, it performed more helicopter rescues in one day than is usually seen throughout the country, over the course of a year. We all had the privilege of watching these brave men and women risk their lives to benefit strangers in need. They told us that they had trained for disasters, but had never anticipated the magnitude of the hurricanes of 2005. They were exhausted, but they did not rest, because they knew that people's lives were in danger. After an open-sea helicopter rescue of the crew of the *Mary Lynn* (in one hundred plus mile-per-hour winds and on forty-foot seas), **Lt. Commander Craig Massello** told us, "We began the duty day just like any other…who would have thought that today would become part of Coast Guard history, and one we would never forget."

I met **Charles Sterling**, Director of Information Technology for Egan Healthcare. Egan had five hundred clients in personal care homes across the region. He effectively pulled together the necessary data so that the health records of his clients could follow them in the evacuation. He has been building on this success and is pursuing new solutions in disaster preparedness. He told me of a nurse on his staff who had the presence of mind to use a chainsaw to cut a path en route to work, so that she could check on her patients. Later, she used the chainsaw to get another patient out of his home before the rising water killed him.

Christopher Porter was one of thousands of American Red Cross workers who helped support the Katrina effort. I met him nearly a year after Katrina. His local Red Cross chapter, American Red Cross of the Susquehanna Valley, Pennsylvania, trained hundreds of people and deployed them to the coast. The chapter also collected donations that were then channeled to the national Red Cross. At this time, the local chapters of the Red Cross are suffering from reduced donations because the public feels that it has already donated funds to the cause. What people don't realize is that the earmarked donations did not support the financial strength of the region, nor do they increase local resources. Despite the current budget challenges, people like Chris are working on innovative ways to help train people like those of you reading this book, with on-site, online or blended approaches to response training (such as first aid, CPR, and the use of AEDs). Now working with InnerLink, he is finding way to make it easy for any school, church, or employer to train staff appropriately, using local Red Cross chapters in a coordinated program.

Michael Hayes, a young New Orleans native and graduate of Loyola University in Chicago, was out of town when the storm hit his hometown. When he returned to his beloved New Orleans, he was shocked by the destruction. He pitched in where he could. He volunteered with Habitat for Humanity, and helped a friend re-open a restaurant. He was given the opportunity to run that restaurant—an opportunity he had long dreamed of—but something about the situation felt wrong to him. He longed for more meaningful work. Then, Habitat for Humanity offered him a job. This is not an optimal place to work if you want to make money, but it offers unparalleled opportunity if you want to make a difference. He took on the St. Bernard Projects Director job and has not looked back. He was involved in getting the "Musicians Village" and two other sites started in New Orleans, and he has worked with Branford Marsalis and Harry Connick, Jr. to make it happen. Each day he watched tens to hundreds of volunteers from around the world slowly rebuild a neighborhood and a heritage. He said, "At freshman orientation they talked about the Jesuit philosophy—the idea that education is good, but needs to be married to service to mankind for the glory of God. I can apply what I have learned and help those around me. I see how those that volunteer with us, or donate, feel, and I know that in many ways, we have been blessed by this storm."

I met Long Beach school board member and NASA scientist **Kirk Sharp** through the homecoming dance our Pennsylvania students organized for students in two Gulf Coast schools. To give his students this memorable experience, he allowed rules to be bent. He was also innovative enough to help keep his school system running even when its core buildings were missing, when half the faculty and students were sleeping on floors or in tents, and when the other half was scattered across the nation. His leadership allowed for the intermingling of Southern and Northern cultures as part of the school experience—a fact that changed the lives of the Pennsylvania volunteers, as our Northerners' stereotype about the South was shattered by the intellect, culture, poise, graciousness, and bravery of those Mississippi parents and teenagers whom we got to know. He and I shared a concern for the health of those returning to their home sites, which were toxic from chemical overrun and biologic waste. He facilitated my discussions of these issues with the top officials in the health department and governor's office. Thank you, Kirk.

These were some of the leaders…the standouts. They showed me successes that can be achieved by those who step up and lead. Thousands of lower-profile successes occurred within each family. Parents and children shared an inner strength they didn't know was there, at a maturity level usually only seen in wise adults. This was not a quick disaster with a clear way out. The successes will continue slowly and in tiny steps, but in aggregate, the locals can be proud of how they learned their lessons. All Americans can be proud of their successes.

3

Evangeline Franklin

I met Vangy at the zoo. After Katrina, the New Orleans zoo was the site of a free clinic that provided primary care, dental care and free eyeglasses to hurricane victims—a necessary service, since the storm had closed area hospitals, doctor's offices and clinics. Because of my experience at a makeshift clinic in New York after the 9/11 disaster, Dr. Dominic Mack, head of Atlanta's Hurricane Recovery Center, invited me to the "zoo clinic," so that I could observe and advise its staff. The place was filled with workers from faith-based organizations who do missionary work around the world in impoverished or war-torn areas.

For a week I watched and helped as five thousand patients arrived for care. People were being fit for eyeglasses that were ground on site, in the pavilion above the reptiles. I saw dentist chairs outdoors, with fifteen simultaneous procedures occurring at once under the sky, or in tents, adjacent to the elephants. A visionary, tired, and intermittently loud woman was in charge, and it was her job to be the doctor for thousands with no budget or facilities. That woman was Vangy. She commanded respect and kept the operation going.

At the end of one day, she burst into tears, saying, "I've got to take care of all these people and I still haven't gotten the stuff from my house. I have no car and no one to help me." I rented a van and we traveled to the shell of her home. We filled it with an assortment of putrefied, wet, muddy items that she could not part with: China, pots and pans, some important papers from her attic, and, pathetically, video tapes of key events in her life. (The water ran out of the tapes as we picked them up. I took these things home with me to try to salvage them, but they were beyond repair.) She showed us her destroyed Steinway piano—a gift from her parents—and the place where her dogs had drowned. We helped her move into a fortress of a new place. It is on the third floor of an ancient warehouse that is made of cement. She was lucky to get it.

I have since learned of her skills as a physician who understands internal medicine, public health, research, insurance, technology and government bureaucracy. I have seen her several times since the zoo, and I have watched her expertly guide her colleagues into the uncertain future. When I told her about my plans for this book, she said, "Let's talk about the successes achieved. There are so many successes, large and small, that have prevented many deaths."

—*Dr. Rob*

It is nearly a year since Hurricane Katrina swept through the Gulf Coast, and my life is finally finding its balance. I spent the past weekend performing the ordinary acts of daily life—washing clothes, putting away dishes, watching the History Channel, reading the paper with a cup of coffee on a Sunday morning, my dog in my lap—and I am grateful that I can appreciate these activities now, more than ever. Although the coffee was made in my pre-Katrina coffee maker, this is a new home, and I am sitting in a new rocking chair that Rob Gillio helped me assemble. After he put together the main parts, I diligently constructed the seat, slat by slat, and reveled in the distraction from painful thoughts and stress.

I rock with a dog that has also survived a torturous eleven-month experience. I found him in a shelter. He was frightened and thin. Now, he has staked out a place in my bed (maybe *his* bed—he sleeps in the middle) and has me trained to respond to his every need. My previous dogs, Mikey and Joey, drowned in the flood that filled my house, and it gives me a sense of balance to know that Frankie and I are recovering from Katrina together.

Since that catastrophic event in the Superdome, I found rest in a hotel room in Dallas, which I shared with three of my New Orleans Health Department (NOHD) colleagues. I worked in shelters and advised Texas mental health workers on how best to approach the tragedy experienced by New Orleanians. (I explained how to engage us with red beans and rice and smoked sausage, not refried beans and tacos, for example. We do not eat that kind of thing.)

On my way back to NO the first time, I attended the funeral of my cousin, who, at the end of a long battle with cancer, had delayed her dying to be with her family in a calm place in College Station. I wailed and sobbed beyond my own belief and spilled out all of the grief that I had inside—grief for her, her girls and for New Orleans.

As Hurricane Rita approached I returned to Dallas. I took back roads and got to Dallas from College Station in a record seven hours. The uncontrollably congested Interstate was a nightmare. I felt a bitter sadness when I learned that even though Texans had bragged that they could handle disasters better than New Orleans, Hurricane Rita created for them the same unexpected chaos we had experienced during Katrina.

I then returned to a black and eerie New Orleans, dotted with National Guard and military security and many, many strangers in my city. I boarded the *Sensation*, a Princess Cruise ship that had been set up as a shelter for government employees who were victims of the storm. There, I shared a dormitory experience with damaged, despondent and—in a couple of cases—suicidal New Orleanians who had lost everything, or nearly everything. On our cruise ship, we had no gambling or drinking or stage shows or beauty parlor—just unlimited amounts of food, safety, security, potable water, storage, a clean bed and laundry facilities. My room became a storage place for my forays into storm-battered houses, and my free laundry bags were filled with foul, water-ravaged clothes that I washed and saved, for some reason—perhaps in order to maintain the idea of control.

My roommate and my friends had to rescue me from my own deepening depression, and return me to the land of the recovering. As Mardi Gras—and the time when we would need to leave the ship—came near, I knew that my fate was not to return to my house. After gutting my home, and my parents' home—an experience that I can only compare to gross anatomic dissection—I made the critical decision to restructure my life on higher ground, on the third floor of a poured concrete building. I took to walking and found a condo within three blocks of the ship. There, with Frankie, I made a home, bit by painful bit. It has since become a wonderful nest, unlike any I had ever had.

I cannot put into words what it is like to have lived through such a traumatic and life-changing experience. I am so fortunate; I have fared so much better than most, and I am extraordinarily grateful for it all. I've also made an enormous amount of progress: Opening clinics, completing Federal Emergency Management (FEMA) project worksheets, redoing budgets, negotiating positions, and overseeing Health Recovery Week and Project Prepare. Although my parents have relocated to far-off Maryland, I know that they are healthy and safe. I have shed my despair, letting go of the past and the sadness, in the knowledge that this new self I've created is capable of much more than I expected. And, in spite of all of this, I've been "recognized" by my true love. I feel that I've absorbed a complete education in a matter of months, and a lifetime of experiences, in less than a year.

Through my experiences, I have learned a few lessons that helped me work my way through to recovery, and that will continue to protect and enrich my life. They are:

- Learn everything—really everything—that comes before you, no matter how boring or seemingly irrelevant. You will be surprised when you use it.
- Practice making decisions, especially at times when there is little to lose from mistakes. Then, learn from your mistakes.
- Do not let fear cloud your judgment. There are times when you will be afraid, but irrationality will lead to poor judgment and a worse outcome.
- Always expect, and be prepared for, situations to change. This will enable you to handle what the future brings, when what it brings is not what you want.
- Believe in yourself, even when you fail. It is more important to learn how to deal with failure than with success.
- Do not be afraid of letting go of the past—and quickly. It is the future that holds all peril and promise, and you must have the skills of a wise warrior to engage it.
- Quickly identify the things that cannot be changed, and proceed to those that can. Learn the consequences of each change you make.

Find your balance—your center, or whatever gives you peace, rest and energy—and return to it, regularly. You will need those moments and memories as you recharge in difficult times. Put things into perspective, and be prepared for the next challenge.

4

Emily Burris

I knew that I had to include stories from Louisiana State University (LSU) after I read "Eye of the Storm," an amazing account of the way that LSU turned its athletic facility into a temporary hospital for a few weeks after the hurricane. As a student leader at LSU, Emily found an opportunity to lead a campus of students willing to help critically ill evacuees. She notes some failures in her school's program, but she also proudly celebrates her part in helping this makeshift, student-staffed facility serve thousands of special-needs patients.

Her generation has been criticized for being materialistic and self-centered. Perhaps it is, but this story, and others that follow, illustrate that many young Americans instead belong to a true Freedom Generation. Armed with the confidence to volunteer, they can, and do, make a valuable contribution to society.

—Dr. Rob

I am a small-town girl from Baton Rouge. In the summer of 2005 I had huge plans for my senior year at Louisiana State University—plans that included tailgating at LSU football games, and participating in activities with my sorority. I was also very involved in campus organizations, and was serving my community as the student government's chief of staff. None of my plans involved taking two weeks out of classes to aid the victims of Hurricane Katrina.

The disaster brought on by this hurricane came as a great surprise, since hurricane season is a way of life in Louisiana. Each year, we LSU students hoped for at least one hurricane to cancel school for a day or two. It never failed that at the last minute, the hurricane would change paths, and we would be glad to have a sunny day off from class. Every hurricane seemed to be "the hurricane that was going to sink New Orleans," and I had begun to regard weather reports with great suspicion.

The weekend Hurricane Katrina was headed for New Orleans began like any other. I relaxed and watched television with friends. I recall looking out the window and watching the wind, without much worry. We only lost power for approximately thirty minutes. The day after the storm, I laid by swimming pool. I remember getting a phone call from the student body vice-president, informing me we would be out of school for the next week and that this hurricane was indeed the real thing.

Later, I was informed that LSU's Pete Maravich Assembly Center and Maddox Fieldhouse was being converted into a facility to house displaced hurricane victims with special needs. Two days after the hurricane, I realized that what had begun as a temporary shelter had become a fully functioning hospital almost overnight. That night I gathered with a group of other students to help organize the shelter.

I recall standing outside of the Maddox Fieldhouse with group of campus student leaders, looking at a building that traditionally holds LSU indoor track meets and athletic practices. I've always heard the phrase "the calm before the storm," and it held true during that moment I stood outside of the makeshift hospital. It was the last moment of peace I would see for the next few months.

Imagine mass chaos, as doctors and medical staff rushed around, volunteers stood, waiting for instruction, and patients waited to go through triage. The scene was like something out of a movie, but at this moment it was our reality, and we had to prepare for the worst. As I think back on our group of students marching into absolute chaos to save the world, I laugh a little. However, that day we walked into a room and we were indeed able to make a difference.

This underscores an important lesson that I learned through this experience. I have always heard claims that today's youth do not volunteer to serve their communities, as if we have no concept of an altruistic act. In a time of crisis I learned that students do step up to the challenge to help their fellow citizens. While this outpouring of selflessness was apparent, it was sadly misused, due to the overabundance of student and community volunteers. I witnessed volunteers being turned away because there was no organizational structure to train them. Hundreds of willing volunteers were told that they were of no use. Their helping hands could have saved lives, but no one knew how to train them, and the medical staff didn't have time to stop what they were doing to direct a volunteer. This lack of structure caused a chaotic atmosphere in a place that was already stressful.

Success finally came when we determined that the most productive way to run the facility was to split the volunteers into two groups: the medical check-in staff, and the organizational staff, which trained the volunteers. Defining these structures helped make the Maddox Fieldhouse organized chaos, rather than the complete chaos we first walked into.

To organize the operation, we first established a definite and easily identifiable registration and check-in procedure. (This prevented confusion within the area where patients were treated.) We also ensured that volunteers, and those in charge of volunteers, were easily identifiable. An organized, well-informed approach to volunteer training can make an enormous difference.

This experience affected all of us greatly. When a disaster such as Hurricane Katrina hits, it affects you mentally and emotionally. When classes finally resumed, the results of Hurricane Katrina made it impossible for us to concentrate on our studies.

Hurricane Katrina altered most plans the student government executive staff had devised for the coming year. We postponed many of our projects to work on a fundraising effort named Mission: Possible. This effort raised over $3 million by sponsoring a viewing of the Arizona State football game in Tiger Stadium; selling Mission: Possible T-shirts; and providing services to visiting students from New Orleans.

Hurricane Katrina helped me become a more appreciative and humble person. Although I come from an area that was affected by the storm I was fortunate, because my family and my house were spared. Watching those around me who lost everything in the storm made me realize that people of all colors, beliefs and backgrounds are affected in the same way by such disasters. I constantly remember the courageous men and women who simply walked up to the Maddox Fieldhouse not knowing exactly what they would find.

Although the days were long, every fourteen-hour shift at the Maddox Fieldhouse truly mattered to those around me and continues to impact my life today. I often think about the women and men whose lives I was able to impact simply by being there for them. I often think of the families I saw reunited, the children playing in the "kids' area," and the first glimpse of hope I saw in the eyes of hurricane survivors.

Because of this experience, I felt proud to be a student of Louisiana State University and a member of the Baton Rouge community. In a situation like Hurricane Katrina which may come round only once in a lifetime, I would encourage others to give as much of themselves as they can, because the opportunity to help mankind, no matter how small or great: will add richness to those that gave and those that received.

5

Rod Schoening

Rod and I met at church. After 9/11, he learned that I needed help treating first responders in New York, and—at a time of serious risk and uncertainty—he showed up and worked hard.

In the months after Katrina, I was not surprised to discover that he had been a volunteer in the early days after the storm. He remains committed to finding ways to improve disaster preparedness and response, and he has changed careers to make that happen.

His first big success in this new endeavor was to enable the Medical Reserves Corps with an electronic medical recordkeeping system called the Health Passport. He also helped the Lutheran Disaster Response use a technology system to facilitate a more efficient, organized and safe response. Now, he manages TeamPrepared, a crisis preventions and response planning program for schools to empower its staff, students, and first responder community.

—Dr. Rob

The day that Katrina hit the coast was just another day and another hydraulic system for me. I was installing a ventilation system in another Amish barn. My job was not especially challenging, but it was different every day, I help people keep their livestock cool—which, in turn, helped make more milk. More milk meant higher sales and the ability to feed one's family and others.

I am a father and a husband. I work hard and go to church. I have suffered personal losses in my life, including the death of a sibling and a parent. I am good with my hands, I've been blessed with a technical mind, and I enjoying seeing a project through from design to implementation. It is gratifying for me to help someone.

I have a military background and I am patriotic, but not to excess. I understand authority, but I get frustrated with it sometimes. I am not as patient as some people—I like to act if action is needed. I help out at church and in the community, and I was one of the volunteers at the New York Police Department's Ground Zero Clinic after 9/11.

I live in Pennsylvania, one thousand miles from the coast. As Katrina approached, I barely noticed the weather reports on the news. What caught my attention a few days later was the constant flow of news images showing people being rescued, or waiting for help. My Amish customers and friends—with their deeply ingrained desire to respond to the needs of their neighbors—were talking about the damage and what could be done. I got time off from my boss, and an Amish friend, Jake Esh, and I loaded up my pickup truck with supplies and a loader with grappler. Then, we drove to the coast, not really knowing where we would end up, or what we would do.

We arrived to find a chaotic situation that was unlike the images we were seeing of New Orleans on the news. The tsunami-like wave that destroyed the Mississippi coast left tens of miles destroyed. All of the debris had either washed out to sea or had moved inland about half a mile, making it almost impossible to get help closer to shore. As we tried to drive farther toward the shore, we approached what used to be the city of Pass Christian. We were stopped by a military roadblock, and were told that no one was permitted on the roads after dark. We were to turn around and go somewhere else. Where?

After talking with officials, an Alderman from Pass Christian drove out to the roadblock and took us to the remains of the city. There, we met another group of volunteers that had arrived earlier in the day. Coincidentally, this group was from our community in Pennsylvania. The leader's name was John Beiler. These volunteers had identified a facility that wasn't flooded, but had suffered severe wind damage. It was on a bit of higher ground in Pass Christian and was called the Gospel Singers camp. They had arranged to fix it up so it could be used as housing for the volunteer workers. We had a place to stay.

As the sun came up we got to see the devastation clearly for the first time. The last time I felt so helpless and scared was at ground Zero while volunteering in New York, after 9/11. Jake and I were sent out with the loader to try and open some pathways along the residential streets. This would help locate remaining victims and allow survivors possible access to anything that might be left of their personal property.

Some other volunteers were doing the same. The abandoned town was eerie, especially with thousands of pieces of clothing hanging in the trees like so many ghosts. "Where are the people, and where were the authorities?" we wondered.

After a long, sixteen-hour day of moving debris, I noticed that Jake had developed a very large blister on one of his feet. The top of the blister was torn open and it was bleeding. I asked him when he last had a tetanus shot. He didn't know if he had ever had one. Jake was sure he would be all right, but I was worried that his open wound might become infected. After all, he had worn a hot, sweaty boot all day, and he had walked on ground that was contaminated with animal and human sewage and remains. The next morning I talked about it with some folks, but they were preoccupied with thoughts of survival, and one man's tetanus shot was far from their minds. Although we were there to help, it was clear that we were also creating a burden on those who now felt obliged to house, feed, and care for us, in addition to the surviving victims.

Later, I made a second trip to the coast. There, I discovered the body of a victim in the rubble, while I was clearing one of the streets. In the military I had worked in an autopsy lab for part of my career, but even that did not prepare me for the emotional hit I would take, seeing a body mangled and interwoven with the rubble pile that was miles long and stories tall. I thought about learning the news of my brother's death, some time before, and I immediately felt the pain that the victim's family would feel when they learned that their missing relative was truly gone.

Upon returning to Pennsylvania, as chance would have it, I discussed this trip with a person from my church who had worked with me in New York, after 9/11. Rob Gillio had traveled with his daughter to the coast to be a part of the rescue and recovery efforts, and we realized that he and I had worked on the same stretch of shoreline. He was one of the workers who served at what is now called Camp Gospel. Well-organized work groups were launched from that place daily.

Rob talked to me about the Lutheran Disaster Response groups that were using his company's software, and said that he needed help with the administration and support of a group that was building another housing facility for volunteers in Biloxi, Mississippi. It was to be built at a church with the same name as ours, Good Shepherd Lutheran Church. Within a week, he had hired me to focus on helping support that operation and, subsequently, to help with school disaster planning and community medical preparedness efforts. I changed careers for less money, but was able to make a difference every day for someone's health or safety.

I learned many lessons during my time in Mississippi. By spontaneously driving to Mississippi to volunteer, I inadvertently burdened a system that barely existed. I did accomplish quite a bit of good, but the destruction was so massive that my tons of rubble movement, in and of itself, were hardly even noticeable. I developed a respect for what FEMA was up against. They must have felt like the Germans did on D-Day, only with a massive frontal attack over miles and miles.

I also learned that one should have never travel to volunteer in a disaster area without first having the basics of their health records with them.

In terms of successes I experienced, my take-action mindset and previously honed skills with the skid loader did make a difference for a few people. The contributions of my neighbors in Lancaster County, Pennsylvania were multiplied as tens of thousands of persons from my area offered days of volunteer work on the coast. There is probably not one church or civic group in our county that has not sent thousands of dollars *and* tens to hundreds of volunteers. Many hands do make a difference, especially when efforts are properly coordinated. Along those lines, the subsequent work I did offsite with technology dramatically impacted the preparation and training of the next wave of volunteers.

My experiences had a personal impact, as well. I found a career niche that has me in the thick of community and personal preparedness. I am challenged to find ways to use knowledge, skills and expert advice to create better planning and response tools and capabilities. I have come to realize that if you have a plan, you can survive—and even thrive—after an emergency. Today, I continue to help the Pennsylvania Lutheran Disaster Response groups as they use our software to organize and prepare for the next disaster. I am also helping a school district create a disaster plan and I am proud of what I am doing.

My life has changed in many ways. Last week, the car my family had just repaired was in an accident, caused by my child. I was very upset because the $1000 repair is not something I had planned for, nor was it the sort of thing my insurance policy would cover. However, I soon realized that this little dent, while significant from my perspective, was sure better than having no car. After all, my child is OK. So am I. I am learning to appreciate that so much of what we think is important is not, after all.

Based on what I've learned I would caution other would-be volunteers not to spontaneously offer to help in a chaotic zone unless they were totally self-contained and would not drain precious few resources from those who really need it. They should also be prepared to change their lives to find a career that is gratifying.

In addition, each individual should plan what to do in an emergency, and should store supplies for themselves and their families. Parents should also ensure that their schools have emergency plans for their kids. Finally, I believe that we all need to realize that possessions are just things—it's people that matter most.

6

Garrett Book

Garrett was a high school kid who was looking for something meaningful to do in his summer before college. He asked if I could find him a way to help the victims of Katrina. I had watched him grow up, and I knew that he was smarter than his years, and interested in many different things. He was considering the seminary; he was also interested in social work and history. I thought he would benefit from his trip to the Gulf Coast.

Tim Brown runs the EMS service in our town. He is also active with Lutheran Disaster response. I met with Tim at an art show held the weekend after Katrina. We discussed setting up a tent city in Biloxi. The same day, I met H.C. Porter, whose photographs would become part of my company's disaster recovery program, Project Recover. Garrett's father, Chuck, was the volunteer project manager pulling that program together. Like father, like son, helping in different ways.

Months later, and just a few weeks before the one-year anniversary of Katrina, Linda Eggbeer, the wife of my college roommate, told me about her experiences at a place called Camp Biloxi, where she stayed while volunteering in the Gulf region. She talked about Tim Brown, and told a story about a young man who played his bagpipes at the tent city they were building. That person was Garrett. It is amazing how small the world is.

—Dr. Rob

I had graduated high school in the spring of 2005, and the summer had been monotonous. I was working forty-hour weeks and preparing for the impoverished college life. At the time, I hadn't the faintest concept of what an "impoverished life" was. I lived with my parents in rural Pennsylvania, where the largest natural disasters included annual droughts and the infrequent blizzard.

On my drive to work one morning, I heard the radio deejay mention a Category One hurricane that was passing over southern Florida. He recounted a phone call with a friend who lived in the path of this hurricane. When the deejay asked his friend why he refused to evacuate, the response was simply, "It's only a Category One." Thinking back to that anecdote often frustrates me; it seems nothing short of a sin against nature to scoff at such a phenomenon. But, of course, no one had the foresight to know what Katrina would become.

I remember watching television on a Sunday night in August, as national news reporters stood like martyrs on hotel balconies along the Gulf Coast. Katrina had swelled to a Category Four and now her solid, slate gray wall was pressing against the beach. Cameramen—perhaps for an adventure, a story to tell or just for a fat paycheck—caught images of roofs peeling off houses and cars rolling down streets like clumps of dust. As Katrina pressed nearer, heroism and professionalism melted away, and all that was left was the sentient instinct to survive. All that was left was mortality. This, naturally, was not revealed until days, weeks and months after the damage had been done.

"Isn't it a shame about all those poor folks in New Orleans?" seemed to be the gist of conversation the next day at work. It filled up potentially silent moments with talk of FEMA and other political jargon that those of us unaffected by the storm had the luxury of discussing. Like all news, though, talk of Hurricane Katrina became old after several weeks. Something changed in me, then. I grew a little, it seems. I felt the need to help in any way that I could.

I knew Dr. Gillio from church and had learned that he was involved, in some way or another, with disaster relief. He had been in New York City four years prior, during the aftermath of 9/11. Shortly after contacting Dr. Gillio, I was introduced to my new boss—a man for whom I'd develop a great admiration in a relatively short period of time. Tim Brown was a gruff man in his late sixties with a short white beard and a white crew cut. He wouldn't hesitate to tell you when you screwed up, and he certainly wouldn't hesitate to give you a pat on the back, either. Experienced in disaster relief, he compared the Gulf Coast to post-WWII Germany.

Tim was the leader. His wife, Ro, accompanied him, offering moral support for Tim, and performing many other important duties. She quickly became moral support for everyone. I would ride down with Leo, a longtime friend of Tim. Leo was a great big Vietnam veteran, a Master Gunnery Sergeant in the Marines and one of the friendliest people I've ever had the pleasure to meet. Along with driving together, we also slept in the same pop-up trailer (As a side note, Leo would wake up early and I would sleep in—sometimes even as late as 6:15, when I would be awakened daily by Leo's booming voice, "Get up Garrett; it's another grand and glorious day…and you're missin' it!")

In the third vehicle were Dennis, George and George. The two Georges, who were both ordained, were often mistaken for each other until Dennis learned that one George was often referred to by his family as "Padre." The name stuck.

They were all loveable characters. Dennis had sold his three businesses in order to spend time working in the Gulf Coast after Katrina. As of December 2005, he had been to the Gulf Coast three times. Never have I met anyone so optimistic, outgoing or eager to laugh as Dennis. George was a sweet, often quiet man, easily approachable and perpetually philosophical. He helped anyone who asked, and would easily take the lead if need be. And then there was Padre. He was a humble man who at first gave the impression that he was rather passive. He quickly became one of the most bold and skilled foremen working on the site, however. In time, one learned that he was never hesitant to give orders, or make a joke.

Our mission was to drive to the Good Shepherd Lutheran Church in Biloxi, Mississippi and build a tent city where future volunteers may reside.

We arrived at the church in Biloxi on November 2. Every house, shop, church and restaurant was capped with bright blue FEMA tarps after being stripped of their roofs. Signs and billboards were bent in half, with no less ease than snapping a toothpick between your fingers. Many windows were missing, and most places of business were closed—two months after the storm had hit. I asked Tim how bad the damage in the surrounding towns was. He simply shook his head with his hands in his pockets. "Oh, so much worse," he said with genuine pain. "So, so much worse." In that moment, I realized that although I had already observed destruction unlike anything most people had ever seen, this was not the full fury of Katrina.

A week of work passed, and when Sunday came, we took an opportunity to see what Tim was talking about. Railroad tracks ran parallel to the shore several blocks north of the beach, and everything south of the tracks was swept into oblivion on the night of August 29. Giant coils of razor wire lay along the tracks, only breaking occasionally at checkpoints so that authorized vehicles could pass to enter the totally devastated area.

Tim was equipped with the appropriate paperwork, whatever that was, and so we drove south of the tracks. What could be recognized as foundations for hotels, motels, souvenir shops, gas stations and restaurants were flattened to the level of surrounding asphalt. As I wandered in delirium where boardwalks and pizza shops once stood—where children had once played, while parents read fiction novels—I felt nothing. I was not angry. I was not sad. There was no remorse. There was no depression. I had absolutely no reaction. As I photographed half empty bottles of red wine and computer keyboards lying on the beach, I tried to cry. Later, we drove along the beach and witnessed miles of dissolution, as Tim told us of people who were killed in the storm. He pointed to their houses, now just splinters of wood in a field of debris. I tried to cry. For days, I couldn't.

But amid the horrific scene of havoc, I saw something else. I saw people from Wisconsin, Minnesota, Michigan, Virginia, Georgia, Alabama, Oregon, California, Colorado, New Mexico, New York, Maine; I saw people who felt they should drop the luxury of their comfortable, daily habits—many of whom were not at all personally affected by the hurricane—travel for miles, and make a pilgrimage to a place in need. It was love. And for two very difficult weeks in my life, that was the greatest lesson that I learned: Love is illogical. That's what's so great about it. Love is illogical.

We established much of the camp before returning to Pennsylvania. Other volunteers replaced us when we left, but Tim made sure to point out that they weren't Alpha Team, and that there would only be one Alpha Team—the first seven volunteers to step foot into what is now known as "Camp Biloxi."

But those were the physical successes. The emotional and interpersonal successes achieved throughout the trip were such that I couldn't possibly begin to sort them out in my mind, let alone write of them. To use the cliché "I'm a better person" is misleading. I'm a *different* person. I'm a more well-rounded person, and I'll always feel a connection to Biloxi. That, I'm thankful for.

I found that to volunteer for anything is an act of love. Acts of love are the greatest acts of all, because love is the most powerful emotion of all. No amount of natural destruction or hate-fueled, man-made destruction can overcome an act of love, because acts of love require sacrifice, and sacrifice requires genuine devotion.

7

Tammy Lee

Tammy was one of several Master of Public Health students from Tulane who worked as interns at the New Orleans Health Department after the storm. All were enthusiastic about their experiences working in the Superdome, and in a health department faced with many challenges. They brought with them as much knowledge as they received.

Tammy's bright sense of humor helped to elevate the gloomy atmosphere among her post-Katrina coworkers. Her willingness to work shoulder-to-shoulder in our stressful environment allowed her the opportunity to become a near-equal contributor during our departmental recovery. Her flexibility in any situation will be the key to her success.

—Dr. Vangy

I am originally from the Lower East Side of Manhattan. For the past fifteen years, I have been traveling around the world. I have worked and lived in New York, Taiwan, Burkina Faso, Alaska, and New Orleans. Of all the places I traveled and lived in, I like Alaska the most. I still consider myself Alaskan because that is where I was the happiest. It is the only place I have been where I was judged by my character, and not the color of my skin. I found peace, myself and God in Alaska.

I earned my bachelor's degree in nutrition and food science from Hunter College (CUNY) School of Health Sciences and my master's degree in international health systems management from Tulane University School of Public Health and Tropical Medicine. This fall, I will return to the School of Health Sciences to start my dietetic internship program to become a registered dietitian.

In my 20s I went to Taiwan four times as an overseas counselor with China Youth Corps (also known as The Love Boat). After graduation, I worked as a dietitian in the Women, Infant, and Children Nutrition (WIC) Supplemental Program at Gouverneur Hospital and then at Saint Vincent's Hospital. From 1998 to 1999, I interned with UNICEF's public relations department. In 1999, I joined the Peace Corps in Burkina Faso as a community health development worker. After my two-year service in the Peace Corps, I was offered a position at Alaska Family Resource Center's WIC Program as a nutrition educator. A year later, I accepted a one-and-a-half-year position as a team nutrition grant coordinator for the State of Alaska's Department of Health and Social Services and Department of Education and Early Development. After Hurricane Katrina, I volunteered at UNICEF, and on my return to Tulane University, I completed an internship at New Orleans Department of Health. Presently, I am working as a WIC consultant at Betances Health Center and a rock climbing consultant at North Meadow Recreation Center.

I enjoy traveling, hiking, indoor rock climbing, and swing dancing. I can spend hours watching action, science fiction, and stupidly funny films.

Hurricane Katrina was not my first disaster. I was in Taiwan in 1996, during Typhoon Herbert, and in New York City on September 11, 2001. On April 2, 2005, my father died and that was my most painful disaster. Almost five months after my father's death, Hurricane Katrina hit New Orleans.

On August 27, the City of New Orleans announced that the hurricane was coming in two days and that evacuation was voluntary. A handful of my friends knew that I did not have a car, so they asked me if I wanted to leave with them. At the time, I did not think that the situation was severe, so I refused the rides. I thought I would be safe because I lived at Tulane University Medical Housing. As I watched the news, I felt the need to evacuate. I called friends who had not yet left. I was very upset when a friend decided not to take any passengers; he evacuated alone. Two other friends whose vehicles were packed, where willing to fit 2 more people into their cars. I decide to find other alternatives. They told me to call then if I was not able to find a way to leave by evening.

At the same time, I was also becoming concerned about another friend, Roberta Manzano. Roberta was new to the country and I was worried that she did not know how serious the situation might be. She is a Brazilian medical doctor who was working on her ophthalmology fellowship at Tulane University Hospital. That afternoon, I called her phone and knocked on her door repeatedly. There was no answer. I remembered that Roberta and eight German medical students (whom I had met the week before) had rented two cars to visit the swamp. They had not yet returned. I was very worried for Roberta because Tulane's evacuation bus was coming to take residents to Jackson, Mississippi. As I was waiting for the bus, Roberta and the Germans came back from food shopping. I told them about the situation and our options. Roberta and I made a promise that we would stick together for duration of the hurricanes.

She rushed to ask the Germans if I could evacuate with them. The Germans said that I could, but only if they had enough room in the car. Later that day, one of the Germans decided to stay in New Orleans, and so that evening, Roberta and I evacuated with the Germans to Houston, Texas. I sat in the middle of the back seat. It was very uncomfortable, but I felt lucky to leave. The passengers in the car were Roberta, Mitja Jandl, Benjamin Berger, Sebastian Winkler and me.

On the way to Houston, the highways and roads were jammed, but we were able to move slowly. We reached Houston around 4:00 or 5:00 in the morning. We were joking about going on a mini-vacation and how much fun we would have at the beach. We thought we were going to have an extended weekend and return to New Orleans in three days.

After the hurricane, we were excited to go back home—that is, until the levees broke. We received a lot of phone calls and emails from friends and family members inquiring about our safety. Many of my friends from different parts of the country (Wasilla, Juneau, New York, Seattle, San Antonio, and Miami) offered me a place to stay. Bruce Botelho, the mayor of Juneau, Alaska and dance instructor of the Juneau International Folk Dancing club, was among them. He is like an uncle to me. I told him I evacuated with four friends. He spoke to Benjamin in German. I was told that they spoke about "the weather."

Every day, we watched *CNN* and woke up to a new surprise. I became addicted to the news. I remember Roberta changing the channel once or twice, so she could watch something more comical. Our group was very sad for a day or two because we thought we had lost everything we owned, either through the hurricane, flood or looting. I remember looking through my small suitcase and wondering why I had brought my ugliest clothes. I felt as though I was in a really bad science fiction movie. I wondered when the zombies would come out.

My group and I felt badly for everyone who was still stuck in New Orleans. We considered ourselves extremely fortunate. Roberta and I prayed every night for the victims of the hurricane. As a group, we started to look out for each other and treated one another like siblings.

For a week and a half, our group moved from one hotel to another. Soon, we started to get evacuee discounts. We also started to share rooms, because our wallets were getting thin. Roberta decided to continue with her studies at Baylor University. I wanted to be with my new friends, close to New Orleans, in case it reopened. I went to Texas A&M and inquired about taking their courses so that I could graduate by the end of the following summer.

At the time, it was not clear how I was going to pay for my fall semester because the hurricane had disrupted my financial aid status. The next day, I emailed and called Columbia University's Admissions Office and they accepted me within three days. I was offered free tuition. I flew to New York within half a week, even though I was not yet mentally ready to leave.

Sometimes, I wish I had stayed in Houston so I would have been closer to my evacuation group. In New York, I was in denial and wanted to continue with my life. I also started to get lonely so I joined one of Redeemer Presbyterian Church's Bible study groups, where I made new friends. During my second month in New York, however, I started to get out of my denial stage. My family, friends, professors and fellow students sympathized with me, but they did not understand what I was going through. I found it helpful to talk to other Tulane evacuees at Columbia. It helped me survive the semester. The Columbia deans and staff members were wonderful. They bended over backwards to help us register and they were very supportive. Our Tulane professors who were in the city visited us at Columbia to make sure we were OK.

In mid-October, we were allowed back into Tulane University Medical Housing. My mother and I went down to New Orleans for four days. I went to my dorm and found that all of my belongings were packed into boxes and placed in a room with other labeled boxes. It was very difficult to find all of my stuff, but I was excited to learn that my computers, thumb drives and passport were not stolen. Afterwards, my mom and I went around the city. The destruction was unbelievable. I felt as though an atomic bomb had exploded into the city and mutated humans were going to come out to eat us alive. We also visited the French Quarter. It did not have the same sexy appeal it held before the hurricane. The streets were smelly. Almost every corner had a refrigerator that had a sign written on it that said, "Do not open, rotten foods inside," or "Please pick me up." A week after coming back to New York, I still felt New Orleans' depression. I was happy that I was in Columbia that semester and was not sure if I was ready to go back to Tulane in the spring.

In mid-November, Mitja and Benjamin came to New York for a visit, so I showed them some of the famous sites. Sometimes my friend Kathleen would visit, and we would all go to a comedy club, to the restaurants, and dancing. We had a lot of fun. Roberta and I tried to meet one another while she was still in the States, but opportunity never arose. This fall, we will attempt to meet one another in New York City. If not, I might have to visit her in Brazil. Another adventure, the Amazon Rainforest, sounds exciting. Though I only knew my evacuation team for a short period of time, I feel as though we are part of a family. I am happy that I was blessed with a really great group of strangers-turned-friends.

I learned some lessons through my experiences. I wish I had brought all my valuables with me when I evacuated. I felt really silly after the levees broke and I could not return to retrieve them until mid-October. For over two months, I worried about my belongings. During that time, it was difficult to predict property control and security of any kind. I realized that not all of my material belongings were as important to me as they were before the hurricane.

Five months after the hurricane, I found out that my credit card number was stolen. The credit card company reversed the charges and gave me a new account number. I even changed my signature. Because of the hurricane, all the credit card companies gave me a three-month grace period to pay my bills. That really helped, because I did not receive most of my financial aid money until the following semester.

In January 2006 I went back to Tulane University for my last semester. My friend Khadija Ellhoujaoui, a fellow International Health student at Tulane and volunteer in the Superdome during the storm, stayed with me until she found a place to live. We immediately became best friends. She told me she used to volunteer for the New Orleans Health Department and believed that they would provide me with a valuable experience. She contacted Dr. Evangeline Franklin from the Health Department and highly recommended me.

At the Health Department, I reviewed the Federal Emergency Management Agency Project Worksheets and worked with the Hurricane Preparedness team in Emergency Medical Records. The work was sometimes intense, but I learned a lot and was treated like a team member. I do not feel I could have had the same experience and opportunity elsewhere.

For my Katrina semester, I went to New York City to continue my courses at Columbia University and Brooklyn Law School, so I would be able to graduate on time. Actually, I graduated a semester early. By working on my capstone in the New Orleans Health Department, I was able to make new connections and learn how the New Orleans Health Department operates.

Before the hurricane, I wanted to apply for Tulane's dietetic internship program, because of their excellent reputation. After the hurricane, I was traumatized, and I felt that it was best that I work on my internship elsewhere.

When I returned to New Orleans, Dr. Elisabeth Anne Glecker, my former professor at Tulane, told me that the Crescent City Cowboys horseback riding club president, Dave Williams, invited me and a fellow Tulane student to join their club. We would ride down the Mardi Gras Parade on horseback. Dave is originally from Tallulah, Louisiana. He is a "black cowboy" and is locally known as the "Horse Whisperer." His superpower is training and taming wild horses. After the hurricane, he was asked by local authorities to rescue farm animals. He fed the animals and rounded up the cattle.

There were other "black cowboys and cowgirls" in the club. As a Chinese-American, I never felt out of place with them. They were very friendly, giving, good-hearted, strong and independent. I do not know how some of these New Orleanians still have a happy face and a heart of gold when they lost their homes and everything they owned. I admire all of them and their strength.

I was very excited to be invited to be in the Mardi Gras Parade, let alone to ride a horse! Club members gave me pointers on how to ride, as I was new to it. I got to ride horseback three times in the parade! The horse I rode on, Cinnamon, was beautiful. (Cinnamon was a bit wild but very careful when children were nearby.) I learned quickly that Cinnamon's kryptonite was carrots. He slowly learned to trust me and come to me. Whenever I would visit Cinnamon, I would clean him, brush him, and pet him. Horseback riding taught me to be at peace with myself and learn to take care of a horse. I still think about that horse from time to time and wish I was in New Orleans to ride him.

I've learned that it is important to have a sense of humor when one is in an awful situation. Smiling and laughing helped keep me going. Roberta and I have an official fantasy evacuation story. I told the story often after I evacuated, but I rarely use it now. Roberta likes to surf and I like to rock climb. The story is that Roberta and I were in the Superdome during the hurricane. After the flood, there was a hole in the roof. I guided Roberta with my rock climbing skill to the roof top. Then we peeled a piece of the roof off. I held on to her, and we surfed away. The story is corny, but I think it is creative.

After Katrina, I became more understanding, patient, and compassionate, especially to those in need. Sometimes, I still have moments when I wonder why some people just can not understand something that is logical and simple. Then, I remember that there are subjects and issues that take me a while to learn and comprehend. Most importantly, I learned more about my own strengths and weaknesses from my Hurricane Katrina experience.

Life is too short to put too much emphasis on material things, because they can be gone in a blink of the eye. It is important to learn to judge a person by character, and not by their credentials, IQ/EQ, wealth, personality, or appearance. Kindness and a good reputation are worth more than money. You will be surprised by how many good people will help you in your time of need. It is important to be happy, comfortable, and honest with ourselves. When we finally face our Maker, we will not be entering into the gates of heaven with our resumes, but with our love.

8

David Ressler

David Ressler is that guy that is always there helping out. When they say "if you want to get something done, ask a busy person", Dave is that person.

He never seeks the spotlight but makes sure the lights are on for others.

When the hurricane came, he felt called and answered the request he heard.

My name is David Ressler and I am employed by Armstrong World Industries, a building products company based in Lancaster, Pa. My functions over 27 years of serving Armstrong have ranged from production associate and production planning to managing a department of 60 employees. During my early years I helped my father with his farming business and also worked for a timber company where I turned many PA hardwoods into lumber for barns (much of which went to Amish barn raisings), houses, furniture, and a large amount of veneer lumber that went to produce burial caskets. I have a beautiful wife, Barbara, 2 children and 4 grandchildren and I am a Christian. I am 52 and still learning what this life is all about. One definite thing I have learned is that we are all in it together.

Katrina hit the Gulf coast on the last weekend of August 2005. It was for most of us, I think, just another hurricane. Nothing to get excited about. A couple of damaged buildings, flooded streets and some displaced civilians. No big deal. We heard it all before. Or had we?

All newscasts began hour-by-hour coverage bringing Katrina right into our living rooms. The way they dramatize things you're not so sure it's as bad as they say, but for some reason this quickly looked like the real deal no matter what or how they reported it. The carnage became evident. The New Orleans Superdome and the 9th Ward housetop rescues started to create the face of a menace that was causing death and destruction right here in the U.S. mainland.

The first confirming sign for me was when the local Mennonite Disaster Relief Agency had continuing news bulletins pleading for monetary gifts and supplies for relief for the tens of thousands of victims. Soon following was the Red Cross making urgent pleas for not just monetary donations but also for volunteers to come to the Gulf Coast to help in any way to relieve the suffering. I discussed with my wife my feeling of being called to serve with the Red Cross but we both decided I should wait to make sure there really was a need. Six days later while coming into the house from working in the yard to check the Penn State football score, there was a newsbreak showing the devastation from Katrina and another plea from the Red Cross for volunteers. I decided right at that moment to go and do what ever I could to help.

On Monday, September 5th, I contacted the Red Cross and volunteered to go to the Gulf Coast. I was required to attended an 8 hour safety course and a another 8 hour training session with the Red Cross "ERV" unit, (Emergency Response Vehicle). After a quick briefing as to when, where, and how I would be serving, I left for Baton Rouge, Louisiana. Since Baton Rouge was far enough north from the coast, they did not suffer significant damage. Upon arriving, I was taken to the Red Cross command center for registration and deployment to the effected areas.

Baton Rouge was at that time in utter chaos with deployed volunteers returning to the command center after retreating from Hurricane Rita. They had no idea where to deploy me or the three to four hundred new volunteers. I took it upon myself to join forces with a young E.M.T. from Lansing, Michigan and started asking where we could best be used to serve. We commandeered a truck and started delivering bulk food supplies from makeshift warehouses to the victim and support staff shelters. We did not come in contact with the psychical devastation right away since the shelters housing the displaced were north of the severe devastation.

One week later I found myself serving with the "Human Services" arm of the Red Cross. One fourth of the New Orleans International Airport was commandeered as a registration post for displaced victims to sign up for monetary relief made possible by the generosity of the American people. It was set up as a

"McDonald's drive-up window" style operation. Victims would drive through with their cars in lines to tell us where they lived before the storm and where they were currently living so that they could be sent a one time relief check for $300.00 for each family member up to 5 members per family.

The third week took me into New Orleans where the devastation was not only the physical damage of homes and property but also spiritual damage as well. I served with a crew of 5 on an E.R.V. delivering hot meals to victims (clients as the Red Cross requested we call them) in the Kenner and New Orleans area. The houses were uninhabitable. Roofs were gone. Some were collapsed. All of them were full of black mold due to the flooding and the intense heat and humidity that followed. Many of the clients that came to the E.R.V. for food had a lost gaze in their eyes. The clients that still had some spiritual well being about them could not thank us enough. One of the beautiful things I remember most about the people was their humility. They did not want hand outs but they knew for now they had no other choice and for most of us who served on the E.R.V.s it was a humbling experience.

I served twenty-two 12-16 hour days. The amount of devastation I witnessed over that short period of time was overwhelming. To write about everything I saw and felt would take the contents of an entire book.

It must be noted that I did not, nor should anyone have gone to that region of devastation as a tourist or sightseer. It should also be noted that we were advised by the Red Cross to be selective when taking pictures so as not to exploit the suffering of the people and their dignity. The cars, the buildings and all the "stuff" could be replaced or repaired. My sole purpose was to help relieve the suffering and do what I could to help start the healing of the people. We were there to repair and build back up the human spirit, our neighbor, who was in terrific need. Nothing more and nothing less.

The lessons were many and continuous. We, the volunteers, had to be very patient but assertive at the same time. When we arrived at Baton Rouge in the beginning of our service, we were ready to jump in and start helping but because of the chaos of the second hurricane they were unable to give us the necessary direction or orders and so we were unable to begin our mission.

Patience was also needed for some of the volunteers who came to help with their hearts in the right place but their mental and emotional state should have told them otherwise.

We learned to be quick with the kind word and a friendly smile. We learned to take time to listen when a client wanted to talk. We quickly learned to be safe and to watch out for each other because of the criminal element present everywhere. We learned for future times to be good soldiers.

The Red Cross has many good and capable people to manage and direct but they need volunteers to the groundwork to get the humanitarian job done. They need people who can take on extra responsibilities when needed and can fill leadership roles if asked. They do not need someone who would enter a relief situation with a commando, take charge attitude. Compassion and humility are essential attributes when serving or leading.

Always listen, ask questions, understand and take small steps to increase your role as you become more comfortable with your small personal position in the big picture of relief.

It was a retrospective look back at what success did we achieve as the plane headed home. Upon arriving back at Harrisburg Airport, I critiqued myself with the simple but complete questions, "Did I make a difference? Did I actually help? Did I make it better for anyone I served or anyone I served with?" My conclusion was yes but with reservations due to the knowledge that there was still so much work to be done. I was there for 22 days but the people of the Gulf Coast would be there for 22 weeks and months and years.

The shelters were supplied. The people were given money for necessities, for psychological reasons of self-worth and to start them back on the road to be self-sufficient as many were before the storm.

International relations were enhanced as I greeted, thanked, and worked with new friends from different countries who came simply to help. A young man from the People's Republic of Russia helped me unload a truckload of food into a shelter where he was serving as assistant site manager. A young man from France helped orchestrate the New Orleans airport project to get the necessary monies to the victims. There were German and Spanish friends helping with the money and food and water distribution. There were no national or geographic boundaries and the unity of the mission overcame the language barriers. We all understood the universal language of helping someone in need.

There was the unity of the human spirit that when seeing another person suffering, reacts with compassion and care and love. The conditions that make for hatred and war come from things outside the human spirit and they certainly were not present in the volunteers I served with.

The experience did not directly affect me professionally. In the farming and manufacturing world, the purpose of making goods or services is money. That is not a bad thing. In the right hands this money is a gift. It allows donations to be given to disaster areas. It allows the freedom of vacation or leave of absence to go somewhere physically to serve the suffering. The one slight effect as I ponder this question is that I have became more aware of the need for patience and understanding of those I work with whether superiors or peers. The manufacturing environment can be rather hostile at times at all levels. An experience such as the one with Katrina, teaches us to do a better job of looking at people and their needs as people who are no different than ourselves. Look at those you work with as if they to, like yourself, can have everything taken away in the twinkling of an eye. An experience like serving in a Katrina disaster area can help you in your working relationship with fellow employees and superiors as it can help you in your relationship with all people, at all times.

I can describe how this affected me personally with one word: "Growth." My mother would take me around the neighborhood where we lived when I was very young with a basket of food and necessities for the elderly who were in need. We would visit and leave the items in the basket and a week or so later visit them again. This was the beginning of a spiritual growth pattern in me but the day-to-day worrying about the cares and needs and desires of this life somewhat stifled that growth. The Katrina experience made me realize that I, we, need to become more active in the call to care for others in need. Kind of like my mother with a basket.

What can I say to others that look to me now for advice after my experiences? Share the peace. Always keep your eyes focused forward. Why? Jesus said the poor and needy will always be with us. Somewhere, sometime, and many times, they will cross our paths. If we look away, or turn away, their suffering will remain and an opportunity for a blessing upon them and us will pass. But, if we keep looking forward, we will see the opportunities to relieve suffering and embrace it.

If one takes on a challenge such as Katrina, there must be critical self-examination as to whether you can experience the suffering and keep emotionally balanced. Question if you can physically withstand the challenge of working continuous 12 and 16-hour days in whatever climate or environment you may find yourself. Question if you have the emotional and spiritual strengths to work, live, and communicate with so many others whose ideas or notions, personal issues and personalities are so different than your own and under very adverse conditions and still be able to stay focused on the task at hand. If not, then do not.

The mission in all reality was one and the same for all the thousands of volunteers from around the world who came together on the Gulf Coast of the United States after nature released it's fury. "Share the peace." For me it was the ability to do it in the name of the one who shares His peace with me every time I take a breath and calls me to do the same with and for others.

Not everyone is called to serve in a Katrina type disaster. But we are called to watch, to care for and see to the needs of others and respond whether it's for a day, a week or a vocation. If we can make a difference, we are obligated to do so.

Todd Beamer was a passenger on flight 97 that crashed in Somerset County, Pa. during the 9/11 crises. Todd was a successful businessman, father and husband. In the biographies that followed, it was not noted that his life was constantly filled with relieving or serving others as his full-time mission in life. But when an opportunity arose for him to serve just for a moment, he did not look or walk away, or cross on the other side of the road. He walked head on into the opportunity to serve and an entire nation was relieved of some of the suffering on that infamous day.

The opportunity to serve may be a moment in time to simply give a hungry person something to eat or send a dollar to an organization helping a multitude of needy. It may be a vocation that serves others in a full-time career or it may be just a minute of time to offer a hug or a shoulder to a neighbor who has lost a loved one. It may be an opportunity like Todd's. Whatever it is for us, if we are called to serve, may we answer it with the faith and the strength and the love of the One who cares for and serves us all, always and forever.

9

Greg Vogt

Greg is an educator who works for NASA. He makes life exciting in science class, offering teachers informative NASA booklets and online supplementary materials designed to help them teach an exciting concept about our world. It was Greg who first explained to me the different types of photographs that can be taken from a spacecraft showing visible light images, images of heat, or movement of the winds and weather.

Under the storm clouds of Katrina, Greg was forced to evacuate, and, like many others, he got frustrated with the rescue and recovery processes. He also shares an important lesson: One cannot depend on the government to solve all of one's problems. In his case, luck and the direction of the wind helped lead him to a successful evacuation.

—Dr. Rob

I have been watching the progress of Tropical Storm Chris on my computer. Until last fall, I didn't pay all much attention to storms brewing in the Atlantic Ocean.

I live a small community just outside Houston, Texas. For most of my fifteen years here, I worked as an educator for NASA Johnson Space Center. In all that time, no hurricane ever crossed the Gulf of Mexico to pass over Houston. We did get a good soaking from Tropical Storm Allison a few years back, but things have been pretty quiet otherwise. Then, Hurricane Katrina struck New Orleans. Houston was missed, but it was a pretty close call. The magnitude of the disaster was stunning, as was the inability of government services to quickly come to the aid of thousands of people in dire need.

As evacuees fled the region, many coming to Houston, we saw constant television newscasts of devastation and of bureaucratic bumbling. Local, state and federal officials found fault with each other while people of Southern Louisiana were suffering.

Here in the Houston area, we began to hear local and state officials talk about their plans if the "big one" should strike here. I remember a local mayor confidently stating that we are better prepared in Texas to deal with disaster and we can benefit from the mistakes made in Louisiana. Somewhere in my brain, I heard a big "uh-oh!" The mayor's statement was put to the test a few weeks later.

Weeks after Katrina, Hurricane Rita began to form and, as it passed into the Gulf, it had the looks of a really bad storm. Five-day projections of its path put Houston squarely in the danger zone. I cut short my attendance at a meeting in Colorado and came back to take care of things in case the storm hit. Local officials were calling for a mandatory evacuation. Getting into town is easy when everybody else is going the opposite way.

With my computer and other appliances unplugged and covered and valuables put away, I joined the exodus. Even before I had gotten home, more than one million people had gone on the road, trying to get away from the storm. Several major highways leading out of town were designated as evacuation routes. Except for the early starters, most people found themselves in bumper-to-bumper traffic. It was reported that Interstate Highway 45, leading toward Dallas, had a hundred-mile-long backup! People spent hours sitting on the routes waiting for openings that enabled them to pull forward a few hundred yards, only to stop again for several more hours.

I left late in the evacuation on another route. About ten miles into the trip, I became a part of another backup that was at least fifty miles long. It was a blistering hot September day and there was not much to do except to drink lots of water, and run the car's air conditioner for a few minutes, then shut the car off. Police, especially in the small communities we passed through, did a good job of herding the evacuees to communities farther north, to become someone else's problem. There were few rest stops, and finding an unoccupied bush along the roadside was a challenge.

I began to track the time we sat in traffic against the time we got to move forward. Openings were appearing every twenty to thirty minutes and once in a while we got to drive a mile or two. By evening, I was still only about thirty miles out of the Houston area—not nearly far enough to be protected from Rita.

Besides the sweltering backup, people started running out of gas. Soon, the highway shoulders were jammed with cars with their hoods up, signaling, "Help!"

I think the most maddening thing to all of us was that only two lanes of our four-lane highway were being used. Police made sure we all stayed on the right side of the road. Once every fifteen minutes or so, an emergency vehicle would come from the opposite direction. Otherwise, the lanes were clear.

Fanning our tempers were radio newscasts reporting on the progress of the storm and the efforts to help stranded motorists. Hour after hour we heard state officials claim that gasoline trucks were coming to help motorists. In what eventually turned out to be a forty-hour period that I spent in my car, I never saw a single gasoline truck.

More news reports stated that police were working on making it possible for us to use the other lanes. We heard a new term: *contraflow,* an evacuation technique in which highway lanes leading into a city are changed into outbound lanes. If the police were working the issue, they were awfully slow. I didn't encounter my first *contraflow* lane until twenty hours after starting out (about three days after the area-wide evacuation began). Another news report came from the Texas governor. He said that the state was doing everything it could to help its people. He then went on to say that he had "warned us to have full gas tanks!" Somehow, he didn't get the point that starting, stopping, and idling hundreds of times used up gas.

Every gas station along the evacuation routes had been sucked dry of gas just hours after the evacuation began. There was no gas to be found within a hundred miles of Houston. People running on empty formed long lines at closed stations. Some stations had hundreds of cars backed up. Rumors fed the drivers' hopes that gas trucks were coming in an hour or two.

The stress of the evacuation wore on all of us, but I was amazed at the behavior of the thousands of drivers around me. People pretty much kept their tempers under control, and they kept in line. I was, however, dismayed by drivers who preferred not to pack their trash. The roadsides were covered by a blizzard of litter.

Eventually, Rita hooked to the east and made landfall near the Texas-Louisiana border. Houston was spared the storm's worst. Had it crossed Houston, hundreds of thousands of people would have been caught out in the open in their cars. The high winds and spin-off tornados could have caused a staggering death toll.

I was in Oklahoma when the "all clear" was called. Local Houston officials sought to organize the return to Houston by asking residents to come back at different times, depending on which area of the city they resided in. I suspect that most of us decided to ignore the plans. I came home by my own route and on my own schedule. A forty-hour trip out took only eight hours coming back.

For weeks after Rita, we heard continuous analysis of how the evacuation went. No government official came out looking good. One local mayor said that when the mandatory evacuation order was given, no one expected so many people to evacuate. (Disaster officials had underestimated the emotional effect Hurricane Katrina had had on Texas residents.) There was also discussion about families that owned several cars. The families took all their cars, leading to further congestion of the evacuation routes. It's hard to guess why officials were surprised by this. More cars meant that more family possessions and supplies could be carried out of the city.

Except for victims of some tragic accidents along the way, almost all of us survived. The lesson that I learned is that we cannot count on the government to respond quickly to a disaster of the magnitude of Hurricane Katrina. Even though Rita barely touched the Houston area, our leaders appeared to be as clueless as the rest of us. Their plans for evacuation and disaster relief were untested and faulty.

It is our responsibility to be prepared to take care of ourselves, and, as much as possible, to look out for the people next to us. Organized help will eventually arrive, but counting on it is a mistake.

Weeks after we had settled back into our lives, I received an email. The subject line read "Shocking Photos of the Damage of Hurricane Rita in Houston." Some joker had taken a picture of his backyard deck with a single plastic lawn chair tipped over. We were lucky.

10

Greg Porter

God's Katrina Kitchen in Pass Christian, Mississippi is a phenomenon in and of itself. It began when Greg Porter took his grill to Mississippi to feed hurricane victims. Now, it is a campus with the largest circus tents you can imagine, serving up two thousand meals a day. It is also a clothing distribution center, supplied by churches and other groups around the world. The volunteer staff is housed in tents and in mini-barns built by Amish volunteers.

Trucks with food and supplies just keep coming, and so do the volunteers. Early on, these volunteers were cared for here while they coped with the difficulties of clearing the debris and, on occasion, of finding a body. Now, they are renovating and rebuilding.

These volunteers come from all walks of life. They are of all denominations of faith, of all races and many nationalities. This place has become the vessel through which a nation's faith-based community can focus its giving, so that it may offer the most efficient and valuable help possible.

—Dr. Rob

I am Greg Porter, a business owner from Western Kentucky with an office in Evansville, Indiana. I attend Christian Fellowship Church in Evansville with my wife, Pam, and son, Matt.

On August 29, while watching the TV coverage of Hurricane Katrina, Pam looked over at me with tears streaming down her face and asked, "You're going to go, aren't you?" My reply was, "I have to."

I got a call from a friend, David Mudd, who had made a trip down the coast to the affected area. He said there was a need for someone to come down to Gulf Port, Mississippi and cook three meals a day for thirty volunteers, for one week. I had experience cooking for three hundred to four hundred people per meal, so I figured that thirty should be no problem.

When we arrived in Gulf Port on September 12, we were informed that homeowners would not be allowed in to Pass Christian to see what remained of their homes until the following day. Local authorities suggested that we take the food and supplies that we had brought with us to Pass Christian and use it to feed and comfort the returning homeowners.

The timing was perfect—only God could have planned it so. We traveled on Menge Avenue until we could go no further. We had reached Highway 90, which runs east to west along the Gulf. We set up our cooking equipment on a concrete medium, in an area of the highway that was still intact, and started cooking hamburgers for lunch. We also had cold water, Gatorade and soft drinks on hand.

Many homeowners came there in need of food and water. We also fed Department of Transportation workers, state police, military personnel, construction workers and Federal Emergency Management (FEMA) employees. The FEMA employees spent much time in the shade of our tents, making contact with the locals as they came for food and supplies. That spot on Highway 90 also became an area of reunion. Many residents found each other there, as there were many local residents missing.

Soon, Hurricane Rita came into the Gulf and we had to find higher ground. While waiting out that storm I met a group of missionaries from Mexico. They asked me to join them on the location that is now home to God's Katrina Kitchen. For a month, I cooked while the missionaries ran a distribution center, offering clothing and supplies to the hurricane victims. We shared everything, and at night we worshiped together.

In the last week of October the missionaries asked if I would continue their distribution center after they returned home. I agreed, and today, the other volunteers and I take work requests for the local families and send volunteers to complete these jobs. We are rebuilding houses and still feeding volunteers and residents.

My volunteer work has changed my life. When I went to Mississippi to volunteer, my wife—who had always been involved in running our business—took over my role in the company, and has continued its growth. I intend to live the rest of my life doing God's work. I will never be the same.

If you feel God is calling you to do a certain kind of work, don't wait until you think you have everything perfectly lined up before you begin. God will provide all you need to do the thing he has called you to do. Sometimes, taking that first step is all you need to do. Don't wait on others—even family. They may not have the same calling as you.

11

The Gillio Family

I am proud of my family. My wife and daughters allowed me to go to a disaster zone when it was still a restricted and dangerous area behind razor wire and guards, infested with sharp debris, murky liquids and malodorous air. They did my tasks around the house and made it possible for me to go without guilt. They understood that people needed help, and they realized we had some of the tools that could rebuild their world. They pitched in, helping to raise money and hammer nails at the Habitat for Humanity House in a Box building event.

At this point in time, members of my family and I have made twelve trips to the Gulf Coast to aid in the recovery from the 2005 hurricanes. There was no family vacation, because our time and money were needed elsewhere this year. They understood this, and were supportive. For this success achieved, I am very proud. They truly have realized that the material things are "Just Stuff" and it is the caring and compassion we share with and for each other that really matters.

At my company, InnerLink, the staff filled in as well. They also worked at a feverish pace over nights, weekends and holidays to support the Medical Reserves Corps, and to get our education projects and our Health Passport ready for use by rescuers, victims and students.

—Dr. Rob

Part One: Beth Gillio

I am the wife of Rob Gillio, founder and CEO of InnerLink. We live with our five daughters in Lancaster, Pennsylvania. Six years ago, Rob changed careers, moving from being a physician in private practice to being a health and safety entrepreneur. As a result, our family has had opportunities to help out in areas of disaster.

Shortly after 9/11, Rob was invited to help run a clinic for the New York Police Department workers at Ground Zero. I later joined him in a capacity of interviewing workers and collecting data. Later, back in Lancaster, I worked with several of our daughters and numerous other volunteers to compile all the information. Since then, our two oldest daughters have done missions trips in various places in the United States and Central America. Most recently, they both have volunteered, on separate occasions, in Mississippi, to do post-Katrina relief work. Rob also has volunteered, both with the girls, and separately.

I describe myself as a volunteer enabler; I tend to the home while others go in a variety of directions to help others. After Hurricane Katrina, Rob suggested I volunteer in Mississippi with our college daughter, Anna (who was making a second trip to Mississippi's God's Katrina Kitchen) and our high school senior, Amy (who was also making her second volunteer trip to the area), but I wasn't sure the timing was right. The trip was planned over Easter, and I felt a responsibility to the other family members to be home for that holiday. However, after thinking about it and looking at the logistics, I agreed to go. I realized that I had the skills to be helpful. I also wanted to see, first-hand, what my daughters and husband had already experienced.

Until this point, my experience of Hurricane Katrina was limited to media accounts and the stories I'd been told by Rob, Anna and Amy. Amy had made a DVD documentary of her first trip and most of what I remembered about Katrina I had learned from her photographs.

What I saw in Mississippi, eight months after her visit, was truly amazing. Much had been cleaned up, but a great deal still needed to be done. Before, where there were piles of debris and mangled houses, I now saw empty spaces. In other areas, nothing had been done. While one house stood, a neighboring house was completely devastated. Plastic bags hung in trees everywhere.

As Amy and I drove along the coast, God's Katrina Kitchen suddenly appeared as an oasis amidst tragedy. It is a camp of large, circus-type tents right on the beach in Pass Christian, Mississippi. A large, wooden cross stands at the entrance. Several of the tents house a large dining area and a commercial-style kitchen, all funded through donations. Others contain food items that have been donated by churches worldwide. Small sheds adjacent to the main tents house some volunteers. Tents and trailers, brought by volunteers, are also on the property. Port-O-Potties and makeshift showers are there for volunteers. The kitchen serves about two thousand meals a day to a variety of people. The food is free and the facility is open to anyone.

The camp faces many challenges. There are so many needs to be met, and it is basically volunteers who fill these needs. It can be challenging to organize these volunteers optimally. Perhaps a coordination of efforts between all of the relief services in the area would be helpful.

At the same time, I saw many successes in action. The volunteers who serve others, and the daily needs they compassionately meet, are inspiring. An example of this mixture of sympathy and competence is a woman from a bordering state who regularly travels to God's Katrina Kitchen with her husband. There, she runs the dining room with efficiency. She and her husband start early in the morning and work until nightfall.

What I noticed most about this volunteer was her ability to make some of the "patrons" feel needed. This was clear in the way she interacted with an elderly man who arrived at the site every day to eat. His wife was deceased and he seemed to be alone in life. She always gave him little jobs to do, because this made him feel needed, and gave him a reason to return each day. Another man, young in years but aged by hardship, was given a job serving beverages. He was drawn to Katrina's Kitchen from many miles away and made it there by bicycle. He was worn down by circumstances, but at the Kitchen, he was given a purpose, as well as a meal.

Another volunteer group was made up of five new and old friends who traveled together and ran the Kitchen, offering a break to one of the longstanding regular volunteers. Several of these women prepare and serve food to needy people in their community. One was a restaurant owner. Another was an interior decorator. At Easter, she brought everything needed to decorate the dining tent with beauty. When they left, another volunteer, who has run summer camp kitchens, showed up and took over.

I witnessed many examples of this willingness to help. When the makeshift showers on the property needed to be cleaned, a youth group showed up and cleaned the showers. None of this was planned; it just happened. Tents and refrigerated trucks were filled with donated food. My two daughters and I were simply cogs in a wheel that, by the grace of God, keeps turning.

I came away from God's Katrina Kitchen gaining more than I gave. I continue to realize that a person doesn't need to be an expert in something to make a difference. As a registered dietitian, I found it interesting that a group of volunteers, many without formal training in food service, could run this large-scale facility, providing three meals a day, seven days a week.

I now understand that a small-scale contribution can make a big impact. Not everyone has the means or ability to travel to disaster areas, but anyone can call a friend to say a kind word, prepare a meal for someone in need, check in on someone who is lonely or sick, organize a day of yard work for someone who is unable to keep up, or even just be that person who enables others to go out and serve.

Part Two: Anna Gillio

I am beginning my junior year at Loyola University in Chicago. I hope to pursue a career in medicine and public health in areas of the world with limited access to healthcare and underdeveloped health infrastructure.

During high school I earned my EMT certification that, coupled with my fluency in Spanish, gave me opportunities to work overseas in medical missions—experiences that intensified my interest in my career choice. I have lived most of my life in Pennsylvania with my parents and four sisters.

As Hurricane Katrina ravaged the Gulf Coast, images of her fury and of the demolition and death she left in her wake flooded televisions and newspapers everywhere. Soon, the images appeared alongside the testimonies of survivors—survivors who, for the most part, expressed a desire to return home.

Many Americans were amazed by the strength and endurance of their Southern neighbors in the face of such devastation and loss, but as they sat in the comfort and safety of their homes, dressed in their dry clothes, enjoying their fresh food and water, and watching their televisions and reading their papers, they often wondered, "Why rebuild?"

When Katrina hit I had just begun my sophomore year of college in Chicago. Due to school I was unable to travel to the Gulf Coast until January. I had been told of a clinic in Pass Christian, Mississippi that needed English/Spanish translators and medical staff, and as a Spanish-speaking EMT, I was eager to volunteer.

Mississippi had not received the extensive press coverage that New Orleans had, so I had no idea how it had fared the storm. From the plane I saw the coast. Countless buildings were stripped down to their foundations, and temporary blue tarp roofs were scattered across the landscape. The view from the ground was even more astounding; even after four months of intensive cleanup and rebuilding efforts, the damage was worse than I ever imagined it would be. Chunks of pavement on Route 90, which runs right along the shore, had been ripped from the roadbed by the surge. Along Route 90 there was a graveyard that was partially unearthed, missing headstones and perhaps a few coffins. Just north, for about a block, the storm had completely demolished everything in its path, leaving only trees and an occasional beam where a building once stood. On a small ridge, a block inland from Route 90, there was a road called Scenic Drive that was once lined with impressive historical mansions, several of which were now not in good enough condition for anything but demolition. Beyond Scenic Drive were remnants of more houses and fields of debris.

With increasing distance away from the shore, the houses' condition gradually improved and the amount of debris slightly decreased. (About a half-mile inland, the homes had sustained wind and water damage, but were standing and, after four months of work, livable.) However, even blocks from the beach, and after months of cleanup, the debris extended for miles up and down the coast and told of infinite loss and suffering. The trees, most of which were missing their leaves, were laden with clothes and plastic bags—eerie, ghostlike, plastic bags—from a nearby Wal-Mart that Katrina had plowed down. On the ground, bricks, roofs, metal beams, trucks, cars, boats, fire hydrants, telephone wires, furniture, fallen trees, even entire houses—all bent and broken—were mixed into piles of other unidentifiable garbage.

For me, the most wrenching scenes were the remains of a child's tricycle and a mangled wheelchair that I saw in the rubble where an expensive apartment complex once stood. The debris even ruined some houses; I saw homes that had been rammed by cars or boats. One had been crushed by a house that had been lifted from its foundation just a few blocks farther south, and had dropped when the surge receded.

I briefly visited New Orleans, and the damage was different but equally unbelievable. Driving through the Ninth Ward was like visiting a ghost town. For miles and miles no other living creature was anywhere in sight. There were no signs left to identify the streets.

Fields of cars, boats and rubble were interrupted by soggy, moldy, rotting houses, and everything was covered in a white, salty film. I found myself asking the questions that I heard so many other people ask before: *How could one even start to correct the damage done? How could anyone ever call the Gulf Coast home again? How could one live at the edge of the ocean or below a levee? Why rebuild?*

On the evening I arrived in Mississippi I learned that the clinic no longer needed me. I didn't know what to do, so the next morning I went to a place called God's Katrina Kitchen, or simply "The Kitchen." The Kitchen was really a whole disaster response complex, located on a lot right on the shore, where I'm told buildings once stood. It had clothing and grocery distribution tents, as well as a cafeteria tent where volunteers and locals alike shared free meals prepared and served entirely by volunteers.

When I entered the cafeteria tent it was full of people, all talking and eating with each other. It was very obvious that this was the town hub; this was my opportunity to network and find someone who could put me to work. I approached some people and explained that I was looking for a group with whom I could volunteer. They introduced me to a man who worked with God's Katrina Kitchen, and he gave me a job at the grocery distribution center.

Although at first I was disappointed at not being able to volunteer at the clinic, working at the distribution center ended up being a blessing. When a week had passed I could hardly bear to leave, and I felt drawn to return. What I experienced was a sense of community among the survivors. They were gracious and kind. Most were patient and all were appreciative. They were also interested in me as a volunteer and wanted to be sure that my needs were being met, even if it meant offering what little they retained after the storm. People looked out for each other and asked about others. They took what they needed and seldom more, despite the fact that they had lost most or all of their personal possessions. They were polite and at times, they asked if they could have something for a neighbor who needed it more. They showed a graciousness I had not seen elsewhere.

When I finally did have the opportunity to return for my second trip in April, I was ecstatic to be able to work at God's Katrina Kitchen again, this time cooking and serving in the actual kitchen. During both trips I ate all my meals at The Kitchen, giving me the opportunity to meet and befriend many people from different places and backgrounds who, even in the aftermath of one of our country's greatest national disasters, were some of the most pleasant, welcoming, accepting and hospitable people I have ever had the pleasure of knowing. The Southern hospitality was contagious and the mood was hopeful, upbeat, caring and progressive.

Every volunteer I met was absolutely wonderful. The first people I met during my first trip to the coast were fellow volunteers at the distribution center. Although they were all friends who had come to the coast together, they all accepted me immediately and treated me as a friend. The man in charge of the grocery distribution was from Canada, and he had quit his job and moved to the coast to volunteer. The man in charge of clothing distribution had come from California, making it his mission to help rebuild the coast and to make the locals' move back home as welcoming and comfortable as possible. Other volunteers that I met during meals at the kitchen had relocated to the coast to help rebuild. In some cases, they brought their families. When my housing fell through, a long-term volunteer I had met invited me to stay with his group, which immediately welcomed me.

Long-and short-term volunteers came and went, but on any given day the license plates of the cars parked at the kitchen during mealtimes showed that people had come from all around the United States, as well as Canada and Mexico. I even met a volunteer from Italy. All of these volunteers were eager and willing to help however and wherever they were needed, at The Kitchen or elsewhere. Although the coast was not their permanent home, they took pride in it and were hospitable to everyone around them, exercising complete devotion to locals and other volunteers alike.

Over time more and more Mississippi residents moved back to their homes, or whatever was left of them. Despite their losses, the locals that I met offered up what little they had left to make sure that I was housed and comfortable. Others learned of my interest in public health as a student, and invited to join them at a meeting that included Governor Hailey Barbour and US Congressman Gene Taylor.

In the kitchen one day I ate with a couple from Long Beach, Mississippi. They had just moved back to their home in April and could not believe the damage that it had sustained. They were friendly and kind, and at the end of our conversation they invited me to visit them the next time I was on the coast, telling me to remember that I always had a family and a home in Long Beach. Perhaps it was their way of saying thanks, or their way of empowering themselves through helping others. Either way, I was overwhelmed by the kindness that I experienced from people who, during what was likely the most stressful time in their lives, showed nothing but gracious humility and humanity toward others.

One woman from Mississippi told me that Katrina was a blessing, calling it "the great equalizer." As we sat in the kitchen tent she pointed out the rich and the poor of the community eating together, for once actually embracing each other's company. They were surrounded by children and adults from all over the globe who had come to serve them, all colors and classes alike. I smiled to myself for a moment, finally understanding the answer to my initial questions.

So, why rebuild? Rebuild because of the community members' renewed respect for each other. Rebuild because of the people who sat in the kitchen tent who, while looking out on their demolished hometown, boasted of its beauty. Rebuild because everywhere one looked were American flags tied to lopsided homes and hand-written inspirational and motivational signs posted next to piles of debris. Rebuild because a young couple from Mississippi chose to celebrate their wedding day in a red-and-white kitchen tent surrounded not by strangers, but by their ever-growing family. Rebuild because this town that lost its floors and walls and roofs will *always* be home, and a feeling that no storm can ever blow away.

Part Three: Amy Gillio

When I hear the word "Katrina," I get a feeling in my heart that I cannot explain. I feel very sobered, very grounded. Waves of memories burn the backs of my eyes and run through my head like a movie I've seen too many times. This is not because I lost my house in the storm and its aftermath. It is not because I lost my school, my job, or my car. It is not because I lost a friend or a family member. I didn't lose anything in Katrina. In fact, for me, Hurricane Katrina was, in some ways, a gift from God.

It sounds strange, and almost heartless, to say this, when so many people suffered in ways I am still unable to imagine. However, I feel I can say with confidence that I have gained more through Katrina than I ever imagined possible from such a crisis, and that is why I feel compelled to tell my story.

Katrina hit our nation as I was entering my senior year of high school in my Pennsylvania hometown. Every day, it seemed, the media bombarded viewers, readers, and listeners with disturbing images, stories, and slander. Chaos, confusion and despair ripped through headlines and it was nearly impossible to imagine the city of New Orleans in the condition that it was. I remember watching a *Primetime* news special that highlighted the political mistakes made in the crucial preparation and response periods. Seeing aerial views of neighborhoods submerged under a brown sea of Katrina's shadow scared me. *What if that was me, waving that flag on that rooftop? They must be so thirsty,* I kept thinking. *They must be so hot. They must be so sad.* I remember seeing images of the New Orleans Superdome. Thousands of men, women, and children were unable to leave and became virtual prisoners of their own city.

Watching these images, I literally became sick to my stomach, and tears streamed down my face. *That could be me,* I kept thinking. Seeing the children hurt me the most, as their wide eyes just gazed into the camera, glazed and confused. As the television showed a woman burying her mother in a makeshift grave on the sidewalk, I had to turn it off. *How could this happen in America?* I felt as though the bubble of my suburban home had burst. There I was, sitting on my couch under a roof that will most likely never be carried away by a storm, in my Abercrombie jeans. My stomach was full; my neighborhood was safe. How could this be fair? The inequality of the entire situation still kills me.

I felt paralyzed. It seemed like any help I could give during such a gargantuan disaster would just be a drop in the bucket. I couldn't think.

Partially due to the tunnel-visioned media coverage, I was unaware that the hurricane damages far exceeded the New Orleans city limits. In reality, of course, it far surpassed the damage on the Louisiana coast. It was not until several weeks later that I even realized that Mississippi was suffering, too.

A group of students from a local high school were traveling to Mississippi to host a homecoming game and dance at Long Beach High School that had suffered greatly in the wrath of Katrina. For weeks following the hurricane, the Pennsylvania students planned a fundraiser that later turned into a full-scale community event. No detail was missed. Over $75,000 poured in from across the county. Local colleges and high schools donated dresses and gowns for the girls in Mississippi to wear to the dance. Football equipment, catered food, a deejay, expensive raffle prizes, decorations, team T-shirts and school supplies were all packed into an eighteen-wheeler that was donated by a local dairy company. Students who attended Long Beach high school's rival school, Pass Christian High School, were also invited to share in the festivities; their school had also been crushed under Katrina's hand.

Forty Pennsylvania students and several teachers and chaperones were chosen through a lottery system to venture the thousand miles to the Mississippi coast, where they would host the event. My father and I were invited to travel with the students and videotape the happenings. This was the first time I had ever even located Mississippi on a map. I packed my bag and drove with Dad to the airport.

I thought this was my opportunity to become un-paralyzed.

We flew into Jackson, Mississippi, which lies roughly two hundred miles north of the Mississippi coast. As we meandered south toward the gulf, the hurricane damages slowly became noticeable. Tarps on roofs speckled the countryside like blue confetti thrown in the wind. Here and there, handmade signs and flags would announce gratitude to the passersby on the interstate. "Thank you volunteers" and "God bless America" cheered us on as we continued our journey.

Eventually, the devastation became more apparent and noticeable. Now, instead of spotting tarps on a handful of houses, it was nearly impossible to find a house without a newly added blue roof. In fact, with every mile we traveled toward the coast, the damage became more and more severe. Stores sat under signs that were mutilated in the heavy winds, and trailer homes lay in pieces on the side of the road. Crews everywhere tackled trees and debris that had become misplaced in the storm.

As we approached Long Beach, I was struck with an awe that left me in silence. It seemed the entire town was leaning. Houses tilted dangerously off of their foundations. Cars, demolished into unidentifiable shapes, sat idly in driveways, almost as if waiting to be driven somewhere. One bent and deformed gas station rooftop rested on top of a severed gas pump that was slanting so far out of the ground, it seemed to be seconds away from spewing gasoline onto the open pavement. I tried to envision what this place looked like during Katrina's whirlwind, for it was nearly impossible for me to imagine a storm this powerful. As the degree of devastation around me grew, I felt a rising feeling of immobility that was almost like a feeling of surrender. What a monster Katrina must have been!

Once on the north end of Long Beach, we met the families that would host us for the nights we spent on the coast. We found that the cliché of "Southern hospitality" could not have been truer, as our host families graciously invited us to set up camp on spare bedroom floors and couches. By this time, Mississippians with standing houses had become very used to putting people up for the night. So many residents had no choice but to open their homes and lives to those who lost more than just a few shingles.

The next day was a whirlwind of preparation and execution of the planned events. The teachers at the school attempted to hold order in their classrooms, but we students were all so eager to meet each other and talk to one another. A happy chaos broke out among us. We tried to keep secrets about what the weekend would hold.

While the homecoming game and dance were bittersweet victories for the Long Beach Bearcats, I can honestly say that I remember little of the event itself. I do, however, have dozens of vivid memories of the smiles under the disco balls and strobe lights, as the Long Beach students, for the first time since Katrina, were able to let loose and forget about the flattened town around them. There were lots of "Wow's" and lots of giggles. We gave away bicycles, DVD and CD players, DVDs, iPods, and dozens of other prizes. The girls looked gorgeous, and the boys looked proud. It was a chance for the students just to feel normal after fourteen weeks of coping with the destruction.

A memory of that weekend that I will forever carry with me is that of Miss Dulcenia. The day following the homecoming dance, a small group of the students accompanied our lead chaperone to Miss Dulcenia's home. That house, which Miss Dulcenia had not visited since the hurricane hit, had been completely submerged during the storm. When we showed up early that November morning, the heat was already creeping up on us slowly. The adults in the group and local volunteers lectured us for what seemed to be hours about safety precautions and sanitary practices we must follow. As the talk of bacteria and mosquitoes droned on, I stared at the single-story yellow cottage, with its white iron porch, and thought: *Her house looks fine.*

When the time came to finally break the seal and open Dulcenia's front door, however, I was met with a sight and a smell that stunned me further into my growing stupor. The entire interior of the house looked as though it had been lifted and shaken relentlessly. A woolly layer of soggy pink insulation coated every surface of the house, making it a breeding ground for slimy, rancid, unidentifiable organisms and cultures. My former experience with "cleaning out a house" meant washing and folding all of my clothes piled high on my bedroom floor. This was unreal. My hands were frozen with my loss of focus. I didn't know where to start, and my head spun as I realized that everything this woman owned was saturated with fourteen-week-old water that had completely filled her home. Everything she owned was buried beneath a sea of pink insulation and brown muck—a color combination that, for the first time in my life, didn't look cute. I felt numb. Anchors, it seemed, were tied to my ankles as I slowly began to gut out the physical remnants of Dulcenia's former life.

Fourteen people worked all day to completely remove the filth from between the home's floor and ceiling. We made huge progress, but I felt as though my brain were disconnected from the rest of my body. My hands were working, my muscles ached from overuse, and my clothes were sopping with sweat. But still, I felt sedentary, as if I couldn't move. Dulcenia's was one of ten thousand homes along the coast that were destroyed. *One ten-thousandth.* Out in the street, as I looked around at Dulcenia's neighborhood, the thought of that statistic seemed to drive nails through my feet. Every single house was in the same condition. *What will her neighbors do if they don't have people to help them? Will more people come to help?* The questions overwhelmed me as I worked among the mountains of debris.

Dulcenia was active in helping us clean out her house. She stood at the door and closely observed the constant flow of traffic in and out of her front door. "Save that! Save that!" she'd say. Sometimes she'd wipe away tears as belongings that could not be salvaged were dumped into the ever-growing piles on the lawn. "Oh, I remember that," she'd chirp, and excitedly tell us anecdotes about the things we transported across the driveway. The adults took pictures of her losses to archive with the insurance company, and spoke to her about her plans for the upcoming months. Toward the end of the day, as a light rain began to fall, she shuddered and confided in a whisper: "This is the first time it's rained since the storm."

As the day came to a close, the previously congested insides were cleared. In a short time her house was completely transformed. The cottage had been stripped away to simply the studs in the walls. Electric cords, pipes, appliances, drywall and light fixtures were all unsalvageable. They looked like bones of a skeleton, sitting awkwardly between four walls. We all felt exhausted and melancholic, and there was little conversation as our team gathered our belongings and piled into the vans to take us back to our hosts' houses. Why did we all feel so helpless, even as we offered our help? My heart stalled with feelings of defeat.

Even though I knew that another group would venture to Dulcenia's house in weeks to come and newly equip and refurbish it, I still returned to the Jackson airport the next day with my hopes low. All I could picture were the other houses in the neighborhood that were partially buried under debris. I pictured Christmastime and birthdays and family dinners in the empty shells of homes that carpet the oceanfront streets. I pictured New Orleans' children, and their cries for help. I pictured six months down the road, when people will most likely slip back into their suburban bubbles and forget that a corner of their nation will be suffering for years to come. I was grateful for the opportunity to help, but I cursed the fact that I was unable to do more at that time.

This sense of disablement accompanied me on other volunteer trips. When working at God's Katrina Kitchen, a massive distribution center in Pass Christian, I met other volunteers from across the nation. I was humbled by their kindness and dedication to their cause. I met a family from Colorado that had traveled in their mobile home to work full-time at the Kitchen; they say they plan to go home "when there is no more need." I met a couple from Arkansas that spends two weeks of every month at the Kitchen to organize the slew of volunteers that finds its way to the center. I met college students who took semesters off to volunteer, and also locals who devoted their days to assisting their neighbors in need. My weeklong efforts seemed insignificant compared to those of my comrades at the Kitchen.

One of the regulars at God's Katrina Kitchen was a local named Max who came to every meal and appreciated each morsel he was given. Always dressed in the same shirt and tie, Max was a lonely soul whose life now revolved around the small tasks he tended to during mealtimes: wiping tables, or folding napkins. One day at dinner, Max talked of how he had lost his sister, his wife, and his mother. He told us he was all alone. I could not hold back my tears as he, with bright eyes, exclaimed, "But it's OK. You're my family now. We're all family now. It's OK."

With that, I could move again.

Max's words are in heart forever, because he taught me something important. Although my feet felt frozen and my arms felt immobile, my mere presence gave one man a reason to get up in the morning. It gave one man a family. It gave one man a sense of joy and a sense of belonging after a life and a disaster that beat him down. The whole time, I had been touching people in ways I didn't understand until Max simply said: *We're all family now.*

This interpersonal connection can only be accredited to Katrina, and that is why I feel she was a strange blessing in my life. I know I have changed. Max isn't just a man in Pass Christian, Mississippi. Max is the reason I feel it is important to use absolutely every gift and opportunity we have been given. We need to exist for each other and think for ourselves. It is through our existence that we touch others; and therefore our lives need to reflect the ability to love that we have been given.

My feelings of paralysis have yet to leave me. The world is full of problems I alone cannot solve. But, I can see that the problems begin to appear smaller as the human spirit grows. The kindness I witnessed on the Mississippi coast is so humbling and so pure, it seems as though I left Mississippi with a heart full of things I didn't have when the plane touched down in Jackson.

I now have the ability to understand that we cannot be paralyzed by our minds and bodies, unable to keep up with the need that exists in our world today. Instead, we need to be grateful for what we are able to do to help, and embrace and accept the responsibilities we have as humans, with hearts and hands are not, in fact, immobile. By giving our best effort and understanding that's all we can do, our family ties with one another become stronger. Our love becomes deeper and our hearts become greater.

We are not paralyzed. We need to be grateful for that fact, and embrace our responsibilities as humans with hearts and hands that are not immobile.

12

Robbin Bertucci

I met Robbin when my daughter Amy and I were in Long Beach as part of the Hurricane Homecoming celebration hosted by Lancaster Pennsylvania's Lampeter-Strasburg High School. She and Amy spent the week together in Mississippi and became fast friends. Just as some need to be helped, others feel a need to help; Robbin and her mother, Debra, graciously facilitated an opportunity for the Pennsylvania students to be a part of the solution.

Robbin is a friendly girl who is brilliant in math. She hopes to do great things—and she will. Her social world at her church was turned upside-down when the church building was destroyed in the storm. To her, that church was a symbol of her community and faith. It also served as the focal point for yearly mission trips to Saltillo, Mexico, where some of the poorest people on Earth live in abject poverty. Robbin had seen human suffering there. Now she has also seen it at home.

—Dr. Rob

Before Hurricane Katrina, my life was pretty normal. It was my senior year of high school, and I had just started a great job as a secretary at my church, St. Thomas. I was very active at church, reading at Mass, and going on mission trips to Saltillo, Mexico. You could pretty much say that I "lived" at St. Thomas.

When Katrina hit on August 29, 2005, I just thought it would be like any other storm. Hurricanes are a normal thing to people on the coast. To tell you the truth, I did not even know about the storm until the day before it hit.

I have learned so many things from Katrina, but the thing that sticks out the most is that the possessions we have really aren't that important. It's the people in our lives that we need to cherish.

After the storm, we were all living without power. We were hand-washing our clothes, and spending more time with our families. As I looked around at all the destruction, I was reminded of the impoverished people I had seen and met in Saltillo. I felt that we Americans were now realizing what it feels like to have virtually nothing.

My church, or "second home," was located on the beach. The storm gutted it completely. This probably hit me harder than anything. I lost my job, my youth minister, and the place where I spent most of my time. I had to realize, as Father Louis says, "The church isn't the building. It is the people who matter."

Our church received many donations for rebuilding, but to me, the most touching act of charity was the three thousand dollars that the people of Saltillo raised for us. It was amazing to know that people with so little could give so much. I believe this was their way of thanking us for helping them.

This year I was extremely blessed to be able to take another trip to Saltillo. I wanted to go back and thank the people there for all they did for us. Because of my experiences in Mexico, I am considering becoming a missionary doctor.

13

Ayanna Buckner

When Katrina hit, Dr. Ayanna Buckner was starting her dream job at Morehouse University School of Medicine, where she was working with some of the nation's leading doctors in the public health field, including Dr. Dominic Mack and former Surgeon General David Satcher. This is about as good as it gets for someone in preventive medicine. At age twenty-nine, she has accomplished more than most have in a lifetime, completing training and certification for internal medicine and preventive medicine, and receiving her masters degree in public health from Yale—all while maintaining an easygoing attitude and take-charge leadership style.

I met her at a New Orleans clinic, where she helped care for the people in her community. She rose to the occasion with hard work, leadership and a vision for use of technology to help improve the safety net in public health.

She admires Dr. Ben Carson, a young African-American neurosurgeon who, as a child growing up in the ghetto, was inspired by a caring and motivated mother, and grew to work at the top of his field at Johns Hopkins University. Brain surgery is difficult, but it is also a high-profile field that is supported with staff equipment and dollars. Ayanna is a lot like him, except what she does in preventive medicine and public health is not brain surgery. It's harder.

—Dr. Rob

In life, we may experience seemingly impossible devastation. We may feel helpless, but there is always something that we each can do. We must seek opportunities to make a difference wherever we can find them, and we must not allow any obstacle to impede this process.

I grew up just outside New Orleans in an area called the West Bank. I commuted to high school and college in New Orleans and fell deeply in love with the city. It always had a whimsical appeal to me, yet many of its people seemed so vulnerable and burdened. Since childhood, I had been aware of the health disparities that plagued my city, and this propelled me to pursue a career in public health and preventive medicine. I left New Orleans for medical school and residency training, but the city remained in my heart. Throughout training, I always remembered New Orleans. It inspired me. I visited home as much as I could, gathering new inspiration to fight health disparities on each trip. In many ways, I felt that I was somehow preparing for a challenge that I could not name, and I wanted to do everything that I could to be ready.

When the first warnings of Katrina were issued, I was packing to move to Atlanta, Georgia. I had accepted a position in the Department of Community Health/Preventive Medicine at Morehouse School of Medicine. Although I tried to stay upbeat, most of my drive was clouded with thoughts of the city. I arrived in Atlanta feeling helpless and disillusioned.

A large part of my family lived in the New Orleans metropolitan area. In the wake of the storm, their fervently expressed expectations resounded in my ears, "We thought you'd surely come here to help. People need medical care here." Didn't they understand that I was just starting a new job? I felt overwhelmed by my desire to help, so I decided to volunteer in Atlanta. However, I found difficulty identifying activities to provide medical care to Katrina evacuees in Atlanta. I enthusiastically sought opportunities with government agencies, nonprofit organizations and churches, but they escaped me. Eventually, I heard of a local church that started a relief center for Hurricane Katrina evacuees and volunteered at their health booth. While there, I saw so many familiar faces from home—some from my grandmother's neighborhood and others who attended my elementary school. Most were eager to share their stories, and I prayed quietly for them. As I searched for meaning in all that had happened over the weeks since the storm, I sensed that both of our journeys had just begun. The relief center closed a week after I started there, but my desire to help was unwavering.

Settling into my new job proved more difficult than I thought. While cities along the Gulf Coast struggled to survive, I felt guilty for the stability of my life. How could I proceed with the normalcy of my own life when the lives of so many people were shattered? I felt so selfish. In February 2006, I began working with the Regional Coordinating Center for Hurricane Response at Morehouse School of Medicine, and my being at Morehouse began to make more sense to me. The center had been funded by the United States Department of Health and Human Services, Office of Minority Health. My first event was participating in New Orleans Health Recovery Week. This activity represented collaboration between the City of New Orleans Health Department, our academic institution, as well as private and volunteer organizations. Various levels of expertise were required to address such a great magnitude of need and to convert the local Audubon Zoo into a medical center. All services were free of charge. They included medical exams for adults and children, prescription medications, counseling services, eye exams, prescription eyeglasses, and dental treatments—including tooth cleaning, fillings, and extractions.

I arrived at the Audubon Zoo unsure of what to expect of the experience. New Orleans' healthcare needs were so profound that people began waiting in line for services hours before we opened the gates of the zoo each morning. Some people arrived as early as 10pm the previous night. Although tired and anxious to be seen by the healthcare volunteers, their spirits were unshakable. Many were displaced New Orleanians who had traveled from cities across the state of Louisiana. Others had relocated to neighboring states but returned for Health Recovery Week because they still lacked healthcare resources.

The start of the week was overwhelming. I had planned to see patients, but was asked to help train volunteers for the laborious intake process. More than 6000 individuals were registered during the week, and the registration station was the major bottle neck when the activity first began. As an undergraduate student at Xavier University of Louisiana, I had helped to coordinate freshman orientation. In my role as the chief student leader on the project, I directed 200 volunteers, as well as planned and managed the weeklong event for several hundred incoming students. It was a phenomenal experience in project management, and I recognized that my expertise could be best used during Health Recovery Week in the patient intake process.

Administrators from the New Orleans Health Department were managing patient intake. They had been running the Health Department with a dramatically reduced staff, and they were exhausted. My presence allowed them a period of respite. My duties changed from volunteer training to coordination of the entire intake station. It was a fast-paced process, requiring immediate mastery of a system in which I had not been involved in planning. Few of the volunteers had any medical training, so I became their reference for interpreting the histories of the patients. I also became the primary translator for the city's growing population of Spanish-speaking patients.

One of my most difficult tasks was deciding and communicating that some services had reached capacity for the day. Though initially disappointed, the usual and immediate response was, "That's okay. Thank you so much for being here. We really appreciate you." The patients were encouraged to return the following morning, and they received first preference for services. Some patients believed so much in the recovery effort that they also returned to volunteer.

In the first few days, it seemed a new obstacle emerged each minute, but the pleasant disposition of the patients triumphed our challenges. Occasional roars of laughter could be heard from the sinuous lines waiting outside the zoo. This signified a quickly developed camaraderie that allowed many patients to overcome their apprehension of having exhausted their supply of potentially life-saving medication weeks prior. Some patients volunteered their places in line to the elderly or to others who appeared to need more immediate attention.

Health Recovery Week was an overwhelming success. $1.2 million in services were provided. More than 2500 patients received dental care. More than 1400 people received optical treatment and screening, and more than 1300 received prescription glasses. The patience, gratitude, and resilience of the patients who participated in New Orleans Health Recovery Week made our twelve to fourteen hour work days more than worthwhile. The week far exceeded my expectations. My heart was warmed by a letter received from a New Orleans native who attended the activity. He wrote,

> "When I found out all of the providers here had volunteered their services totally free of charge, I could not see how such an elaborate clinic could be brought to a remote location such as our city zoo. I felt grateful for this wonderful outpouring of help...after I began getting treated for the various needs I had, I really was deeply touched. I have never experienced such a great charity (love) from total strangers."

What did we learn?

Collaboration is paramount. No one entity can address the needs of people affected by Hurricanes Katrina and Rita alone. Local, state, and federal branches of the government must work with private groups such as corporations and educational institutions. Through this diverse resource base, we can assemble the expertise that is needed to produce successful events and interventions.

We all have more power and influence to effect change than we realize. The most important step is often recognizing the opportunity to make a difference. Most of us felt unprepared to meet the needs of those affected by Hurricanes Katrina and Rita, but we were unable to ignore the deep call that we felt to help. We relied on all of our experiences. My medical training was only a small part of my preparation for New Orleans Health Recovery Week. My leadership experience and years of community work with vulnerable populations heavily influenced my contribution to the activity. It is paramount that we acknowledge our strengths and determine how they can be most effectively used.

We achieved successes in collaboration and healing. Through New Orleans Health Recovery Week, we brought healing not only to the physical aspects of the people of New Orleans but also to their spirits.

The collaborations established during Health Recovery Week built strong relationships for future collaborations. The Regional Coordinating Center for Hurricane Response at Morehouse School of Medicine and the City of New Orleans Health Department have since worked together on Project Prepare, an initiative to help New Orleanians prepare for hurricane season through organizing their health information into an electronic medical record.

Early in my life, I identified my interest in addressing health disparities in New Orleans. My mentors stopped short of discouraging me from embarking on such a challenging task. "That place is like a third world country," said one of my mentors who was raised in South America and now works for a United States health agency. Another mentor who directs a large metropolitan health department referred to New Orleans as "a public health nightmare." These statements were made regarding New Orleans *prior* to the storm. Health Recovery Week provided an opportunity for me to address health disparities in New Orleans as I never imagined I could at this point in my career. Consequently, I have reexamined my research and outreach priorities. I have decided to devote a larger portion of my time to public health interventions similar to those I have done in New Orleans.

Healing is fostered through helping others. New Orleans Health Recovery Week was a very personal event for me. I had watched the news footage of the hurricanes ravishing the Gulf Coast. I was devastated, and Health Recovery Week helped me to work through my pain. On television, I saw the despair of my home and my people. In person, Health Recovery Week allowed me to witness the hope of the people of my city, and it allowed me to personally contribute to sustaining that hope. At times, I am overwhelmed by the task of eliminating health disparities in this great country. However, my experience at Health Recovery Week fuels my desire to work tirelessly toward achieving this goal.

I came away with this thoughts I would like to share.

In life, we may experience seemingly impossible devastation. We may feel helpless, but there is always something that we each can do. We must seek opportunities to make a difference wherever we can find them, and we must not allow any obstacle to impede this process.

14

Kevin Stephens

Kevin U. Stephens, Sr., has served as the director for the City of New Orleans Health Department since 2002. He is a board-certified obstetrician and gynecologist, a Fellow of the American College of Obstetricians and Gynecologists and an attorney. His most recent presentation was to the Institute of Medicine in October 2005. He has written newspaper articles on various health issues. He currently serves on many federal, state and local boards, and he travels frequently to speaking engagements across the country. He attended Southern University, where he obtained a BS in Physics; LSU Medical Center, where he obtained a MD; and Loyola Law School, where he obtained a JD. He is an active participant in many local, state and national societies.

He is also a good guy. He took time out of his busy schedule to discuss with me his ideas about how to empower patients and caregivers to acquire the knowledge, skills and resources they need to secure their own health and safety. He had an inside view of this disaster and has had to deal with widespread illness and death in his city, a significant loss of budget and staff; and the lost infrastructure of his city's healthcare delivery system for the poor. He took the time to work on this essay because he cares about the people here and in other cities, and he is hopeful that this essay will spark others in similar situations to better plan for the future. His essay was written in early August, 2006, as he and his staff prepared for another hurricane that forecasters warned was then heading their way.

—Dr. Rob

Overview: Heroes and Leadership

During the time surrounding Hurricane Katrina, many heroes placed individual safety and security at risk for the benefit of others. As director of the Health Department for the City of New Orleans, I was able to observe these courageous people as they made tough decisions. I worked with staff in adverse conditions. I met experts from around the world who came to lend their expertise, and I now have the opportunity to help paint a vision for the future. In this essay I will try to give an example of the successes we experienced in the year that followed the storm and to demonstrate the ebbs and tides of genuine leadership, as I saw it, from my vantage point at the New Orleans Health Department.

Many of us frequently underestimate the potential long-term impact of our common everyday decisions. We often lose sight of the way that individuals can make differences in the lives of many. We may lose the ability to gage when and where it is appropriate to sacrifice our personal or corporate duties for the needs of the greater good of society. What we need to realize is that heroes do not go out looking for opportunities to express their greatness. Rather, *the real heroes are ordinary people who chose to make a difference by regularly contributing small but important advances in the lives of many, one person and one day at a time.*

This smaller-scale approach is not high profile. In fact, many people who have had great impact are not recognized, because they are in a supportive role. The administrators and volunteers who set up New Orleans' Superdome—an arena that was set up as a shelter of last resort for those residents who could not get out of the city, for a variety of reasons—are a case in point.

In the Superdome, many lives were saved from the widespread flooding and destruction of Katrina. Despite the problems of no air conditioning, no electricity and no working bathrooms, the lives of most of the residents in this shelter facility would have been in jeopardy if the Superdome had not been made available to them. Furthermore, the devotion and commitment of the many individuals who staffed the facility have been predominately ignored by the media and residents.

Throughout the rescue and recovery process, it became very apparent that the solutions to most problems are not very far away. Time and time again, miracles have happened not by magic or hocus-pocus, but by appropriately using resources within our reach. The take-home message here is that *leadership skills start with ordinary people who extend help to the less fortunate.* Furthermore, the resources required to meet these needs are usually within arm's reach. This is the real story of the heroes involved behind the scenes, under the cup of Hurricane Katrina.

Finding mentors: Mayor C. Ray Nagin and Deputy Secretary of Health Claude Allen

As the city of New Orleans responded to Hurricane Katrina, I was privileged to watch persons I admire learn to deal effectively with this unprecedented situation. One of those persons was New Orleans' mayor, C. Ray Nagin.

The first time I met C. Ray Nagin was when he came to New Orleans' Lakeland Hospital to speak with our physicians. A rather tall, slim and well-dressed man, he was quite a distinctive individual. His speech was succinct and precise, and was delivered in a heart-warming cadence. He was well received by my peers and colleagues, partly because he took the time to communicate with them about their concerns with city government. Based on the perceptions at the reception, we were convinced that he would win.

He did, indeed. Not long after the election, we met at Stryers, a contemporary local establishment on St. Charles Avenue that was designed for the young professional crowd. Mayor Nagin exuded a unique blend of charisma and humility. When I offered him my proposal—a three-paged outline on what I thought needed to be done to revamp our health department—he listened quite intensely and asked a few focused questions about the logistical process. Clearly, he had a tremendous capacity to size up the subject matter, and us, quite quickly. This *ability to size up situations, circumstances and individuals quickly and accurately* is an important skill for any leader.

When I concluded my explanation, he asked me to consider being the health director in his administration. I declined the position, since I was quite content in my current position as an obstetrician in private practice and as the State of Louisiana Medical Director for Women's Health in the Office of Public Health.

About two weeks later I met with Claude Allen, the newly appointed Deputy Secretary of Health for Tommy Thompson, and Ricky Keyes, who had received a doctorate in health information from Tulane University, and who had worked with me on several projects in the private sector. Claude was very bright and articulate. When I told him that I had been offered a position to assist the mayor in building healthcare in New Orleans, he said that I should undertake this huge challenge, because he felt that I could make a positive change. He—and Mayor Nagin—have since become mentors of a sort to my wife Cheryll and I.

This brings up two imperative points about leadership. The first one is that *it is important to have role models/mentors who are accessible.* It is critical for leaders to look to individuals they admire for inspiration and motivation, and yet many of us pick role models who are out of reach. The take-home message is to *focus on the positive attributes of the mentor, as well as the process and pathway leading to, and arising from, the development of these attributes.* Interaction with the mentor is the key ingredient—without this you are just a member of your mentor's fan club. The second point is that *great leaders select and nurture apprentices liberally.* Many times, leaders get caught up looking in the mirror and suffer the fate of the mythical Narcissus, who fell so in love with his own reflection in a pool of water that he drowned in his attempt to embrace his image. As individuals develop their leadership skills, a good gage of true leadership is to count how many individuals they are mentoring.

After careful consideration of the landscape, Cheryll and I concluded that I should accept the position and work with the mayor to help reform healthcare in New Orleans. (This brings up another component in leadership. It is paramount that *spouses and significant others are included in the decision tree of major career shifts.* Taking a position in the public sector had a tremendous impact not only on me, but also on my entire family. During the evacuation in Hurricane Katrina, my family had to leave New Orleans without me, and I was away for many days during this disaster.) With the mayor's help, I was ready to move the status of New Orleans' healthcare forward.

Lessons Learned through Hurricanes Isiadore and Lilly: Preparation, Reward and Consequences

It took me a few weeks to rearrange my schedule to accommodate the new position. Within the first month, I was faced with Hurricane Isidore. There was a fairly brief schematic of a disaster plan on a shelf in my office. After reviewing it, we prepared the Superdome by pre-staging supplies and arranging an area for housing the special-needs patients.

Clearly, the disaster plan needed to be revised and clarified, but fortunately, only twenty-seven patients showed up at the Dome, and we were able to keep the system rolling. We stayed in the Superdome overnight and returned the Dome to the management organization the following day.

The challenges we faced ranged from the simple to the complex. For example, many of the patients had respiratory problems, and lying on cots without any elevation of the head proved to be very problematic. Several patients forgot their medications and did not know the name of the medication that they were taking. We had to be prepared to administer a variety of medications without the benefit of the previous medical record.

The lessons we learned from Isidore turned out to be invaluable in the formulation and execution of our plan for Hurricane Katrina. We allowed one caregiver for each patient, and we found out very quickly that the caregivers frequently had medical problems—sometimes greater than that of the special-needs patient. On the flip side, we learned a lot from the caregivers. One patient was having some respiratory distress. We were considering transporting him to the hospital, until his caregiver said that this was his usual state. She promptly jarred him in the back and cleared up his airway. Another learning opportunity arrived when an infant in our care started to have seizures, and we had to call for an ambulance to transport the child to the hospital in the middle of the storm.

Right after Isidore, Hurricane Lilly hit the coast of Louisiana. It did not cause the tremendous damage that was anticipated. Nevertheless, it was imperative that the hurricane plan be revised. We assembled our team and worked diligently to develop and revise the plan, in light of the recent problems we encountered.

When the hurricane season ended, we went back to our original agenda of improving the health status of the citizens in New Orleans. In a particularly productive meeting with Mayor Nagin and Claude, I was able to watch these two leaders interact. It was apparent that both had a clear vision of their respective roles and responsibilities, and they also showed respect for each other. The important point I learned was that *in order to be successful, a leader must possess the ability to create and sustain complementary relationships, which must be grounded with mutual trust and respect.*

After this meeting, Claude told me the secret of his successes as a CEO. He said, "Every six weeks, I fired someone who needed to be fired; I promoted someone who deserved it; and I painted the lobby." Several months later, it dawned on me that this was the ultimate model in action.

I was able to put this philosophy to work when, after about six months as the health director, I had begun to receive anonymous letters that were mailed to the Mayor, city council members and the chief administrative officer. They were quite ugly and embarrassing, and they included distasteful and offensive comments made about several of the staff members. It was very obvious that the author was someone inside the Health Department.

Remembering Claude's strategy of rewarding deserving employees, I instituted a recognition program that included a picture and write-up in the quarterly newsletter. I also made special attempts to praise deserving employees. It so happened that the employee of the year was my secretary, because she had really gone the extra mile to help me get acclimated to the department, when I entered the post.

Despite all of these efforts, I kept receiving hateful letters and each time, they became more insulting and evil. I came to realize that I was only doing half of Byron's formula: I needed to enact consequences. In the next staff meeting, I told the staff that things were not getting done and I was left no choice but to tighten up the management. While going through this process, I noticed a letter on my secretary's desk, addressed to the prior health director.

I came to learn that these letters were linked in some way to my secretary. After consulting with the city attorney, I issued a letter to her, giving her a notice of a pre-termination hearing. (She admitted to subverting the mail in the past.) Eventually, I transferred her to a position more in line with her credentials. After this incident, the hate letters stopped! Byron's formula was effective. The lesson for leaders is that to be effective, they need to *judicially and prudently use a combination of rewards and consequences.*

Mayor Nagin, who also exercised this formula, later exemplified further lessons in leadership. In the beginning of his administration, I was invited to his weekly staff meetings. These meetings were very informative, with each person reporting critical and pertinent agenda items. This highlights an important leadership skill: *leading a group session*. The Mayor was excellent in keeping the staff members focused while extracting the important action items. He was also able to *ascertain the group dynamics prior to engaging in dialog not related to my area of expertise*. The mayor correctly shot over the boat, so to speak, before sinking it, giving the captain adequate warning to change directions. He was able to adapt to the situation quickly, and make changes, as needed.

Engaging Amtrak

One of Patrick Hall's best attributes is his ability to root out important information. Patrick can find a needle in a haystack. When given a jump-start, he nets valuable information. We were in the process of planning a project called "Getting New Orleans Healthy". Patrick, Dr. Evangeline Franklin, and other team members worked long and hard hours putting this document together. Patrick was assigned by me to work on the hurricane evacuation plan.

This had the workings of a great plan. While I was at a conference in Washington DC, I met with Larry Beard in the Lobby of the Renaissance Hotel. After a fairly quick conversation, we decided that it was a good idea to incorporate Amtrak into our evacuation plan. It should be mentioned that recent Amtrak accidents in Louisiana had resulted in several deaths, and so Amtrak would have benefited from such an arrangement.

We outlined a plan in which train cars would be transported to New Orleans ninety-six hours prior to landfall. We would be given a train to transport citizens from New Orleans to Hammond. This train would go back and forth until the tracks were shut off by the levee board. He asked that we make the arrangements to transport the citizens from the train station to a shelter, and told us that Amtrak had successfully provided a train for evacuation during Hurricane Ivan. He mentioned that Amtrak was willing to expand and extend this operation for other community members. I could see that Larry had a great ability to quickly assess the circumstances and develop an executable action plan.

On my way back from this trip, I emailed Patrick the outline of this plan. Patrick took this outline and ran full speed ahead, in his usual manner. After a few weeks, Ms. Josie Harper of Amtrak rode a train down to New Orleans to meet with us. Ms. Harper handled herself quite graciously. She appeared to have a genuine desire to help us develop a workable plan where both parties would benefit from this arrangement.

Based on train schedules and track ownership, we decided that the best place to unload would be Hammond, Louisiana—a little more than an hour's ride from New Orleans. (Apparently, because a different owner claimed the track into Baton Rouge from New Orleans, it would have taken much more time to get to Baton Rouge.) We worked through the logistics of loading and unloading the train, including issues relating to the needs of wheelchair-bound citizens. By the way she escorted a slightly confused, elderly passenger off the train, I could tell that Ms. Harper had a down-to-earth and caring manner. The important point I learned from her is that you must *get out of the ivory tower and touch people.* In order to plan for citizens to ride the train, we had to go to the station and get on the train.

Once the background work was completed, Patrick called around Hammond to find out what facilities were available. Within days, he had assembled the maps, floor plans and contact information needed to develop this component of the plan. Patrick contacted the University Student Center to determine the seating capacity and layout of the shelter. Additionally, he contacted the director and Local Office of Emergency Preparedness (OEP) administrator to start the process of developing a plan. After contacting the administrator, they concluded that the area was very small and that they were not ready to accept Orleans Parish residents. They recommended that we look at other facilities to accommodate the residents. Patrick got right on this and searched for schools and churches to determine where we could find a facility large enough to house our residents.

The next step was to contact the Pete Maverick Center and Centraplex in Baton Rouge. Patrick speedily retrieved similar profiles of the two large buildings, with the corresponding capacity estimates and floor plans. Again, several obstacles were encountered, making a simple concept difficult to execute.

It became very quickly apparent that we could not develop this plan in the health department alone; we had to include OEP. We contacted OEP Director Chief Matthews and arranged a meeting. He had recently been appointed as the head of this office by the Mayor. Chief Matthews brought with him Louis Murrulo, his deputy. Louis was originally a police officer who had transferred to the OEP office to help coordinate both departments. He had a rather slender frame, with a stern face that projected confidence.

In the coming weeks we also held a series of meetings with the Red Cross, the New Orleans school board and the Regional Transit Authority. As the plan evolved, a tremendous amount of enthusiasm developed. The school board agreed to allow ten schools, which were strategically located throughout the city, to be used as staging areas. It also offered to staff a hundred school buses to transport citizens to Baton Rouge, with one trailer bus in the convoy, and it agreed to supply a mechanical bus for road repairs. To my surprise, the school board also decided to attempt to repair the school buses if they broke down on the road. The collaborative spirit in the group was tremendous. I really felt that everyone had the best interest of the citizens at heart. Ultimately, Amtrak agreed to stage the largest train available.

The final link in the development of this plan was to arrange for shelters in the different sites for the residents. The Office of Emergency Preparedness had set up a meeting with the shelter task force to arrange for the coordination of the evacuation. This was the last critical link needed in this plan. Unfortunately, Katrina arrived before the meeting could be arranged. Time ran out, and storms wait for no one. We failed to complete the plan and implementation on time.

Successes Achieved and Lessons Learned

In the midst of the failures, the planning process produced solutions that, in part, were still useful, even though the grand plan was not finished. In other words, the initial planning made it easier for us to improvise where the plan was not complete, or when systems failed.

It should be noted that the RTA advanced the ten para-transit vans for the special-needs patients. With these vans, we were able to transport over 450 patients to Baton Rouge. This cut the number of people in the Superdome to 600. We transported the most critical patients first. This allowed my staff left in the Dome time to tend to more patients. We also used the staging areas to load the buses and bring citizens to the dome. Even though the plan as a whole could not be executed, several components of that plan *were* indeed implemented, preventing many deaths.

Through this experience we learned that one should not wait before planning. Don't let politics and influence get in the way of what is needed for your citizens. Get the best people on your team and empower them to act quickly. Have the leadership to improvise when things go wrong.

The Aftermath: My Staff

All of my staff members either lost their homes to Katrina, or they opened their homes to those who had nowhere to go. They all lost someone in their families, or knew someone who had died. They all experienced major challenges as a result of Katrina, and had little inner resources left to help others. Most lost all of their possessions. Most of their families moved away to live with relatives and to find jobs. Many are still traveling long distances on weekends to see their spouses or children. All are dealing with massive financial issues, such as arguing with an insurance company that claims that water damage is different from wind damage and that the particular damage they have is not covered. All are waiting for repairmen to start or finish the endless repairs.

Those who have come back to work are faced with a diminished staff, as many of their coworkers have moved away, and some have died. The budget for the department is uncertain and the tax base for the city has dwindled. Referring a client for services is not so easy when the service provider agencies or practitioners have the same issues; they have also died or have moved away, and have lost their offices, their records, and their jobs.

For me, a major success achieved in this disaster was to learn from my team and to be motivated by them. Those who stayed are my heroes. They are doing what they can with the resources they have. They give a hundred percent at work, and then do the same for their families and neighbors when they get "home". I am learning about professionalism and commitment and I am enriched by those I direct.

My Vision

In some ways I now have the opportunity to help shape the future. A major success achieved is to be in the right place at the right time. My city and office have seen the best and brightest descend upon the city—some for their own purposes, but most, to help. These people are from government, academia, faith-based and other non-governmental organizations and corporate environments. I can learn from them and can use their resources to create a new paradigm in healthcare.

A blank slate has been handed to me with tools I can use to make a new safety net for the poor's healthcare needs. Going beyond the safety net, we must to build an infrastructure of technology, methods, institutions and people to create a new system that promotes health, wellness and safety. Information can be used to cut errors, reduce waste and coordinate care on a day-to-day basis, and this information can be a lifesaving tool in emergency and disaster situations. Innovators such as Dr. Rob [Gillio] have been anticipating this opportunity and have ideas and tools to share. Leaders with resources, such as Secretary of Health and Human Services Mark Leavitt, have a similar vision and can bend the rules for the state of Louisiana so that we may try to create a new future for wellness, not just a safety net for illness and injury. I believe the major successes to be achieved are in process right now, as I am surrounded by motivated staff and a dedicated set of outsiders who wish to assist as we create the future for New Orleans.

15

John Tardibuono

"Dr. John T," as the kids know him, is a school psychologist who has long offered valuable help to severely traumatized children. Along with Dr. Lark Eshleman, he is a key founder of the Institute for Children and Families, which offers services to children who have suffered unimaginable hardships. Some have seen a parent shot in front of them; some have lost a parent when the World Trade Center towers fell. The needs are great.

When I developed the concept for a public awareness campaign to make the nation's schools aware of the potential for post-traumatic stress disorder, I turned to John and Lark for advice. John, who is arguably one of the nation's experts in using the computer for psychological intervention, has contributed significantly to Project Recover, an online program for use in schools and clinics.

John is hard on himself for not doing more face-to-face work with Katrina victims, but I can attest that he came down to the hurricane-swept area with me, slept in the car with the mosquitoes, and visited and talked and listened—and listened and listened—to victims, mental health professionals and volunteers at God's Katrina Kitchen, and at Long Beach High School, in New Orleans. He then created tools that teachers, counselors and psychologists can use across the nation to help the children affected by Katrina and other tragedies. His essay helps explain the ongoing help that traumatized individuals need. Sadly, his expertise along with others at the office at InnerLink, are now being used with the members of our community after a tragic shooting at an Amish school just on the edge of our town.

—Dr. Rob

In the days, weeks and months following the devastation from the hurricanes of 2005, countless volunteers offered their time and expertise to the people living in the Gulf Coast region. Their good works included everything from rescuing people to serving food, to gutting and rebuilding houses and mending wounds, both physical and psychological. Regrettably for me, I was not among them. Six months after the hurricane season I spent many days in the region listening to people's stories and struggling to find a way that I could be helpful.

Many of my earliest childhood memories consist of being startled awake by the shrill sirens of police cars and fire engines every night; of always being alert to the likelihood of a vicious fight breaking out on the playground basketball court; of the sights and smells of the intermingling Italian, black and Hispanic ghettos. In those days, and in that environment, I was one of the lucky ones. I got to go to college, and then, to graduate school.

Shortly after beginning my graduate work, luck was again with me. I was given the opportunity to help write some of the first-ever college courses to be delivered via a computer. It's hard to believe that was forty years ago! In 1966 I had no idea how powerful a teaching method the computer, combined with the Internet, would become.

Almost forty years later, when the rains and wind and flooding and trauma arrived in the Gulf region in late summer of 2005, I was offered the opportunity to prepare materials about the impact of trauma—materials that would be shared with young people impacted by the hurricanes in the Gulf region. I was energized. I had found a way to lend a hand, not only to those very much in need of mental health information and services, but especially to those most in need, who, by virtue of the circumstances of poverty, would be least likely to recognize the symptoms of their distress, and to seek and get the help they need and deserve.

With this opportunity I sensed that I would have the chance to share the lessons I had learned from growing up in the streets of the Bronx (the chronic presence of danger and violence can lead to defensive behaviors which appear to be offensive); working for thirty-five years as a school and clinical psychologist in an inner-city school system (when children, teens and their parents can tell their story in a safe, accepting situation, the causes of mental pain and behavioral difficulties can be identified and healed); having been trained and certified as a mental health first responder (that there are marvelous strategies and tools, based on recent understanding of neurological changes caused by trauma, which can relieve suffering); and from reading the findings of inspired, creative researchers as well as the ideas of innovative clinicians.

Trauma reduces or eliminates one's ability to trust. Those who suffer individual trauma, abuse, abandonment, life-threatening injuries or illness may very well lose the ability to trust people; individuals who suffer community-wide trauma may lose the ability to trust people *and* the world.

Traumatized individuals experience varying degrees of hyper-alertness or numbing, avoidance of people and/or places, and re-living or re-experiencing both the emotions and images of the traumatic event. Victim shame can inhibit disclosure. Hyper-alertness or numbing often leads to building personal walls of protection, using the building blocks of aggressive acting out or passive turning inward. Walls erected to protect themselves from further harm also keep out those who are trying to help.

Unfortunately, in under-staffed clinics and over-crowded schools, too often what's seen and heard becomes identified as symptoms, and diagnostic or delinquent labels are assigned to traumatized individuals. In too many instances, such labels mask the underlying, neurologically based changes that can help explain *why* those behaviors were likely to occur.

Sensitive, responsible use of the Internet can provide opportunities to help traumatized individuals learn to recognize significant symptoms; to realize that there are reasons for those symptoms; to understand that help is available; and to know that they can be empowered to play a major part in their own healing.

Using similar e-education technology, teachers everywhere can learn the three R's of trauma: *Recognize* (the symptoms), *Respond* (make classroom and instructional modifications) and *Refer* (to know when a student's needs require clinical, rather than educational attention).

The knowledge, skills and techniques to relieve mental and emotional suffering, and to greatly reduce destructive behaviors, are out there. Project Recover is one of those tools. Effective use of those tools requires learning to see, rather than look, and to listen, rather than hear.

The high-profile circumstances of the destruction of lives and property brought about by the hurricanes of 2005 should help raise the national consciousness of the potentially long-lasting emotional impact of trauma. Images, and their related emotions, continue to be "locked" in the brains of many individuals who were present at Columbine or 9/11, or who were abused, or who live in war-torn cities—or in any neighborhood where violence is a daily occurrence.

What we've learned through everyone's efforts to help relieve the psychological, mental and emotional impact of trauma resulting from the Hurricanes of 2005 can be added to the wealth of clinical and bio-scientific knowledge already in place. We are learning to help guide victims to move from surviving, to recovery, to thriving. The next challenge is to share ways that individuals, from preschoolers to senior citizens, can prepare mentally and emotionally to greatly reduce the likelihood that the next traumatic event will be traumatizing…it does not have to be!

16

Kirk Sharp

Kirk and I met near a tree called the Friendship Oak. Tradition holds that those who meet under this five-hundred-year-old tree will be friends for life. Kirk is an open-minded school board president who, with teacher Matt Cooper, brought together students from two different parts of the country for a unique and moving experience.

One group of students needed help. They were depressed and their worlds in Long Beach, Mississippi had been turned upside down by the hurricanes. They had lost everything. Their senior year experience was a disaster, and there were no plans for the all-important Homecoming. The other group of students had seen the post-Katrina suffering on television, and they felt a deep need to help. They suffered survivors' guilt, watching the disaster on TV, living in a beautiful location and enjoying the spoils of a pampered senior year.

The two groups came together near the Friendship Oak. For a weekend, the students of Long Beach and their parents experienced a Camelot-like escape. For their part, the students of Pennsylvania had given generously but received much more in return.

When it ended, students from Long Beach's Madrigal group sang "It's a Wonderful World" under the tree. After the song, the crowd broke up and left quietly. My eyes met with Kirk's, but we were both too choked up to speak—and we weren't the only ones.

Since then, Kirk and I have joined forces to find ways to be sure that the citizens returning to their home sites are safe. Kirk is back at work at NASA, and continues his service in the Air Force Reserves. After work, he deals with the school, community and family recovery issues.

Kirk was, and is, successful in dealing with each stage of this disaster, and his success is a model for the nation. One school was totally destroyed. The remaining school was operational soon after the storm, and it ran on two shifts to accommodate all of the students. The school leadership was brilliant. The students whose school had survived the storm hosted students from the school that had not. This offered a sense of empowerment to all of the students. A second, temporary school facility now is in place, so the double shifts have stopped, but the camaraderie continues.

Kirk inspired me to write this book, primarily to motivate people in leadership to do a better job. He can teach that lesson by example.

—Dr. Rob

Just one month shy of the first anniversary of Hurricane Katrina's impact on my home, extended family, community, and infrastructure, the memories blur about just how we managed to muddle through. It's hard to communicate the magnitude of this life-altering event. It's difficult, now, to realize that within the span of nine hours, one in four of our friends became homeless; nearly all our childhood landmarks and favorite stores, restaurants, and parks were wiped clean away; and the familiar hustle and bustle of a vibrant community turned into an eerily quiet place. You don't just get up and begin to put the pieces of your life back together in this type of situation. First, you deal with the shock and disbelief, and you worry about how your loved ones are making it. You become overwhelmed with emotion as you discover each surviving friend and loved one who ventured to ride out the storm. Then, you take stock in the fact that you now find yourself seeking food, shelter and security. For those of us who had the additional burden of leadership, it became time to allow experience, training and fortitude to take hold. You see, we had a community to put back together, and friends and loved ones to help.

I live in the little, cozy town of Long Beach, centered on the sugar sand beach coastline of Mississippi. I am an electrical engineer with NASA's John C. Stennis Space Center in my civilian job, a reserve Air Force Colonel in my military job, and a school board president in the last year of a five-year appointment. Nothing in my experiences came close to prepare me for the challenges that Hurricane Katrina presented.

Although I vividly recall the elements of my training taking hold in the hours after the storm, it was impossible to be fully prepared for the magnitude of this disaster. Some of my contemporaries had experienced Hurricane Camille in 1969, but all of them said that Katrina was "no Camille".

It's also important to note that we planned and prepared for the event of a hurricane, but *we really didn't plan to recover from the hurricane's aftermath.* Honestly, you don't practice recovery, you practice preparation, and I'll admit that after the largest national disaster hit us—well, for a short while, we didn't know quite where to start. It didn't take long for us all to realize that stabilizing ourselves, individually, was the first priority. We then began to help each of our neighbors, and then turned to help our community. Chronicles of the devastation in photos and articles clearly detail our plight, and much has been said about what help did or did not come, but I'd like to focus on presenting observations about events that I believe represent important lessons learned.

Evacuation

This storm's quick pace from Category One to Category Five left many with less than forty-eight hours to prepare property and pack valuables for evacuation. Main interstate and highway arteries became clogged quickly in the preceding two days before landfall as lower Louisiana, New Orleans and the entire Mississippi Gulf Coast attempted to use these roadways at the same time. Interestingly, I evacuated just eighteen hours prior to landfall and, after spending nearly an hour on Interstate 10 moving about two miles eastbound, I reached some familiar back roads leading north. *Having a map, I negotiated back roads,* moving toward my final destination some 250 miles due north. I made the trip with only about an hour and a half delay. Friends and some family members evacuated a hundred miles north, staying on the main artery. It took nearly three times longer than it would have under normal circumstances.

Those who lost everything share a common regret of having lost one or more family heirlooms. It's not practical to leave things packed at home that you'd require for an evacuation, but in the quick pace of evacuation it sure would be nice to *have a checklist of what you value most, and the materials necessary to pack it.* Some family members share stories of how hasty packing caused damage to their most precious possessions, even during their attempts to keep them safe.

Response Phase: Days 1–6

We, the victims, found ourselves stepping up to be the first relief workers. We were followed by our first responders. Next, volunteer first responders, faith-based organizations, institutional relief teams and government assistance efforts arrived. Our community members took pride in helping each other and pooling resources and supplies. *City leaders, local companies and the community began clearing debris from roadways in the first hours after the wind and water subsided,* allowing first responders access to the most severely damaged areas. Some areas were not cleared for days because of the magnitude of the debris, and so local residents and the first responders walked through up to eighteen-foot-high debris piles to search for our neighbors in need.

After the initial search for survivors and victims, the first priority became getting water, food and shelter for everyone. Most important for immediate survival was having *enough food and water on hand for at least three to four days.* Having survived eight previous hurricanes, we had upwards of fifteen days' rations stored throughout our wind-battered home. Since communication was gone, "sneaker net" and word of mouth were important means of keeping people aware of distribution locations and relief assistance. Fuel for vehicles and generators was precious—and, in the first weeks, scarce—and those who had vehicles used them to get water, food and shelter for their families.

Within days, church-based relief workers set up food distribution locations and makeshift medical clinics. Because of the magnitude of the disaster, relief workers were in short supply in the first weeks of our recovery. Water and ice were distributed from centralized locations, first by community members such as our high school football players, and later, by the ever-growing number of relief workers.

One indelibly etched memory about the first days following Katrina concerned the heat. The eerie calm and clear skies of the first three weeks in September resulted in temperatures of over one hundred degrees that slowed our recovery pace and threatened our people and relief workers. *Knowledge of how to handle high heat stress environments was key.* Drinking plenty of water, taking frequent breaks, and knowing the warning signs of heat exhaustion were a must. It was not unusual to see neighbors checking on each other.

I recall taking about a week to patch my roof, remove trees, repair decking and replace missing shingles and ridge vents with plastic. Throughout the project, I was careful not to work between 9 AM and 4:00 PM, to avoid the intense heat of the mid-day sun.

Recovery Phase: Day 7

Getting the schools operational was also key to our sustained recovery effort. The familiar surroundings of teachers and friends provided a sense of normalcy for the children in our battered community. Parents knew their kids were safe and being cared for while the adults dealt with the devastation surrounding them. School activities, even pieced together, provided distractions from the stress of the astonishing situations. In one instance, the first local football game between rival schools turned into an incredibly emotional event as players, fans and officials met at mid-field after the game and shared their common bond as survivors. There were also emotional reunions, as some who had returned to the area very recently attended the game.

Our school district suffered varying degrees of damage to each of the five schools. Luckily, the high school, middle school, and two of our three elementary schools suffered repairable roof and internal damage. One elementary school, located near the coastline, was so severely damaged that it was not usable. Quick actions by our administrators allowed the schools to open just one month after the hurricane hit. The students from the destroyed elementary school shared one of the repaired schools for roughly three months, in split shifts. Each school stayed together as a group, allowing the children to keep friends and teachers they knew before the storm. Nearly four months after their school was destroyed, these children enjoyed a normal day of school in modular buildings placed on the north end of the elementary school they had previously shared.

An additional consideration for the administration was that most of these kids were from the school district that received the most destruction. Over eighty percent had lost their homes and virtually all their possessions. The magnitude of need within this group became a concern for many of us. We took special care to see to their whole-body health. *What proved most beneficial for these children was the re-connection with their friends and their teachers.* Their first day of school looked more like a visit to grandma's house than a return to school. I've never seen so much hugging at a school—kids with kids, and kids with teachers.

Another observation I would like to share deals with the large scale of aid, offered by so many. Aid is important, but the wrong aid at the wrong place and at the wrong time creates a drain on sparse resources during such a recovery effort. *Assistance coordination is vital!* Where possible, we used volunteers from our community to clarify the type of aid necessary, and where it was needed. In some instances, organizations sent advance parties for the purpose of targeting aid relief efforts. They first coordinated with local leaders and then determined the best way to deliver the aid for maximum benefit.

During the process of a long recovery, it is also important to understand that the required type of aid changes rather dramatically. As our community recovered it was again able to care for itself in some areas, allowing efforts to shift toward other needs. At first food, water, and shelter were the main considerations. We also began to need medical care, unskilled labor, construction labor and basic supplies. Then, our needs changed to such things as school supplies, building materials, skilled workers, and planners. Communicating the need and getting it to the right place at the right time was a very gratifying job, and daily coordinators would see the product of their labors manifest into hope for so many.

My last observation relates to the long process of recovery. Having come to the Gulf Coast as a youth in 1970, I recall that it took nearly twenty years before the entire impact of Hurricane Camille was removed from the landscape. Now, a year after Katrina, only a small number of places have been rebuilt. Everywhere you look, open fields and slabs provide reminders of the forceful nature of her eye wall. Many of us have more than one family living under one roof, and the real heroes are those who have volunteered to care for others for the long haul.

My wife and sixteen-year-old daughter are heroes of mine. My daughter gave up her bedroom to live for a year in shared, cramped quarters, never once complaining or thinking of herself. She showed compassion and love to her aunt and uncle and their thirteen-year-old son, who had lost everything in the storm, and even helped welcome their two dogs into our home. My wife, a nurse for a local hospital, stayed during Katrina to coordinate hospital activities, enduring a flooded first floor and severely damaged fourth floor. Her courage and commitment didn't end with the repair of her hospital; it extended to the way she compassionately helped family and friends. As my wife and daughter exemplify, having *abundant patience* is the key. Being prepared for the long-term nature of recovery helps keep in perspective all that must be done, and the relatively slow pace that it will all happen.

These few observations loom large in my mind. The experiences of this natural disaster indicate clearly that no one organization, government or community is able to provide immediate help to so many in need, and so it is up to the individual, and the local community, to do everything possible to help. Personal responsibility, preparation, and recovery planning are things each of us can do to ensure we survive—and over time, prosper—after such incredible devastation.

17

King Lam[1]

King Lam is the sort of man who combines hard work with a love of life. A devoted instructor of tai chi, he drew upon these ancient teachings and practices to weather the challenges of the storm and its aftermath. He had also devoted his life to teaching social studies in the New Orleans Public School system—a system that, after Katrina, had no students to teach for nearly a year. King was laid off. He needed a dentist. He had not had good dental coverage before the storm and would have had to pay cash for the services he received. But now, he had lost his tai chi studio, his home, his job, and his health benefits. He came to Health Recovery week as a patient and as a volunteer.

—Dr. Vangy

I was born and raised in New Orleans, and I've seen my share of hurricanes. Even so, I was not prepared for the destruction that Hurricane Katrina swept into my city. Persistence, hope, the help of others and the principles of tai chi helped me survive the ravages of Hurricane Katrina—and to persevere, with faith in the future, through the long recovery process.

I was at my girlfriend's home on August 29, 2005, when Katrina hit New Orleans. As the storm traveled through our area, we saw—and felt—the high winds knock down surrounding trees and blow part of the roof off of her home. We lost electricity and phone communication, and had to rely on a battery-powered radio to learn the news.

The next day, I was amazed by what I saw in my neighborhood. Houses, cars, trees, telephone polls and places of business were all destroyed, to varying degrees.

1. Co-authored by MaryAlice Bitts

Looters had broken into businesses and had taken what they could. The city had lapsed into a state of lawlessness, and the losses were immense.

The worst destruction, however, was yet to come. The 17th Street and Industrial Canal levees had broken, and we watched as water slowly rose over uptown. Early the next morning, I went outside and found that the floodwaters had reached to my knees. I quickly realized that the flood was spreading throughout the city, and that we had to evacuate.

I woke my girlfriend and her daughter, and together, we drove through the rising waters, toward Memphis, Tennessee. At one point, we got stuck in the mud, but through the help of fellow New Orleanians, we escaped to the Interstate. We were on our way to safety. After a week in Memphis, we moved on to Houston, where my girlfriend secured a job and a place to stay.

Although we were safe in Houston, I worried about the people I had known in New Orleans. It was very stressful. I kept in touch with loved ones via email, and I discovered that my siblings had evacuated to sites all around the country: Delaware, Texas, Nevada, New Mexico and Arkansas. I also checked the Internet for news about New Orleans. What I read was unsettling.

I learned that the death count from Katrina was 1,073, and that my mother, sister and brother's house had been flooded by six feet of water. Another sister had one side of her house torn off by winds, and the house was later condemned. My home had about $4,500 in damages, and my girlfriend, mother, two sisters and one brother lost their homes entirely. Some of my students had eight to ten feet of water in their homes. I later learned that the school system in which I had worked as a social studies teacher for twenty-nine years had closed. I was instantly unemployed, along with 7,500 other school personnel, and countless other workers throughout the city. It was a very difficult time.

Back home, the city was in shambles. Mayor Nagin did not allow us to return to New Orleans for a month after the disaster, and he enforced curfews on the city. The National Guard was brought in to restore a sense of order to the town and stop the looting and lawlessness brought on by the flood.

In the months that followed, returning residents were often unable to find work. A low student and patient population meant that university professors and doctors had to be dismissed from their posts. Others who wanted to work were met with frustration when, like me, they came back to find their places of business ruined by the waters. Seventy-five percent of the roof on my tai chi business' building had been ripped off in the storm. The water damaged the carpet, ceiling, equipment, and wooden floor, racking up an estimated $41,000 in damages, and putting me out of business—a business I had built up over the past twenty-two years.

To make matters worse, we returning business owners soon discovered that our customer base was gone. A few months after the storm, the population of New Orleans—which was about 500,000 before the storm—had been reduced to approximately 100,000.

There was good reason for this. By the end of 2005 there was still no electricity, sewage, gas or water service in some areas of the city, and many traffic lights were still out of order. Debris—including refrigerators, stocked with rotting food, and left to the side of the road for pickup—was still littering the streets.

These conditions came as a shock when I came back to New Orleans. Many of my students have still not returned. Reconstruction is a challenge, because contractors who can rebuild a roof, rewire electricity, reinstall plumbing, and perform all of the tasks necessary to rebuild a city are in short supply.

As we struggle to deal with the aftermath of Katrina, we all must learn to cope, and move forward. For me, this meant keeping in mind the four principles of tai chi: *Peng, Lu, Ji* and *An*. These principles helped me remain flexible to life changes; to gather up the energy I needed in time of crisis; and to center myself, despite the chaos surrounding me. I also worked to reestablish my yin *Yin* and *Yang*—my balance—to become reconnected with the earth, my community and myself. I believe that we all must employ each of these age-old techniques not only to survive, but also to live healthfully and well, regardless of the challenges we face.

It is not easy to rebuild a life in the face of such widespread devastation. There are many jobs to do, and it will take some time to reconstruct New Orleans. It will not happen overnight, but I believe that it will happen. With hard work and faith, we can rebuild our city, so that it enters a new renaissance.

If we each work to center ourselves, and to give back to the community at large, we will see this renaissance take shape. We will discover the hope that still lives in New Orleans.

18

Vivian Ward

Viv is a teacher in Mississippi's Long Beach High School. She loves her students and passionately believes that their education should be meaningful. An excellent public speaker, she spends hours coaching speech and drama. Her students admire her, and I can see why. Although she is a survivor of Hurricane Katrina, she doesn't label herself as a victim, but instead, as one who has experienced two hurricanes firsthand.

—*Dr. Rob*

On August 16 1969 I had everything, materially, that any slightly spoiled only child would want. This changed the next day, when my world was shattered by Hurricane Camille—a storm that ravaged the Mississippi Gulf Coast, leaving only debris in her wake. You may wonder how this thirty-seven-year-old memory relates to my current life. In truth, what happened then helped prepare me for the devastation I now see each day. You see, thirty-seven years ago I witnessed the most extreme sacrifice that I had ever seen, and it prepared me for the valleys that were to come in my life.

My dad was an amazing, self-taught guitarist. I vividly remember sitting at his feet as he played his beautiful Gretsch guitar and sang in his deep bass voice. The music filled our home. When Hurricane Camille hit——destroying his guitar, along with our other possessions—that music was silenced. Weeks after the storm, sitting in a tiny government-issued trailer, I realized that I needed music in our home again. There was no money for food or clothing, let alone something as superfluous as a musical instrument, but that didn't stop an obstinate ten-year-old from praying. Every night I would kneel beside my tiny bed in that miniscule trailer and would pray that God would send my Dad a guitar for Christmas.

On Christmas morning, 1969, I awoke with a sense of apprehension. *Had God heard my prayers? Had He answered?* I still remember a ten-year-old's elation when I found a guitar case under the tree. I struggled with its fasteners and then held up the guitar for both of my parents to behold. But then I stopped short, puzzled. My guitar was much too tiny for my father's giant fingers and hands. I looked into my parents' tear-filled eyes and began to comprehend: The guitar was for me, so that I did not need to depend on anyone else to make music for me.

Now, fast-forward thirty-seven years. The Mississippi Gulf Coast lies in ruins again, because of another catastrophic storm. Once again I am a witness to great sacrifice—this time, in the form of thousands of people from all walks of life coming to help us get our lives back. As in the Bible, I am witnessing strangers washing our feet. The faces I have seen, young and old, all communicate the same message to me: They say that we are loved and important. I would like to thank those volunteers for their good works, and I promise that those good works will be passed on.

19

Chris Porter

After Hurricane Katrina Chris Porter and Karol Sessums, who make up the art studio H. C. Porter Gallery, dropped everything and focused on creating ways to raise national awareness about the need to help with the area's recovery. They put together the "Backyards and Beyond...Helping Mississippi Recover" art collection and exhibition. These visionaries are photo and audio journalists and artists, helping as they can. This essay is used by permission of the artist.

—*Dr. Rob*

My artwork has always depicted Mississippians in ordinary settings: A woman shelling peas, a man sitting on his front porch, kids playing basketball. My characters have gone through a volume of experiences, which become evident the moment the viewer responds to the piece. The initial impetus for my subject matter began in December of 1992 in my studio neighborhood, a primarily black community in Jackson, Mississippi. My characters exert their own power and individuality with such force. Each piece is charged with an emotion. As a social realist, the emotionality, enthusiasm and Southern experience in my work has always been apparent, a documentary of history and Mississippi culture.

As a professional independent artist I exhibit my work in galleries and fine art festivals throughout the nation. Although Mississippi is my home, I travel extensively throughout the year with my partner and marketing manager, Karole Sessums. Scheduled to be exhibiting my work in Lancaster, PA on Labor Day weekend of 2005, I wasn't even in Mississippi when Hurricane Katrina slammed our state. On August 28, 2005, "Evacuation Sunday," we found ourselves traveling north, surrounded by Louisiana and South Mississippi license plates fleeing to Memphis, Tennessee for a haven and higher ground. By Monday afternoon, August 29, along with the nation, we were all consumed with the images and stories. Thankfully our family had survived, yet everything, it seemed, had become disposable. The Gulf Coast of Mississippi was leveled. Houses, where they existed, had become tablets for gruesome messages from search teams. Remaining walls had frantic messages of survival: "All O.K. Thank You Jesus!" The coast was a wasteland of debris and death. New Orleans was, for the moment, spared the horrible blow. However, by Monday night, New Orleans was flooding. Thousands of people, predominately black, were clinging to life and family on their rooftops and in their attics, trapped by rising floodwaters.

Meanwhile, Karole and I were traveling with our mental images, but without access to real pictures...until we stopped Tuesday night to get a hotel room. What we saw was this tenuous hold on life, as people plucked from their rooftops were set "free" at flooded interstate exit ramps to be abandoned for days with no aid...no food, water or toilet facilities. Our nation and the world watched as Americans were left to their own squalor. The plans of the city were bleakly flawed. The media continued to gain access, but our government couldn't! We were seeing people on television who had had their support system shaved away. America has a good heart when it cares to see. We were seeing and we were shocked! However, we were in Pennsylvania to work. We still had to set up my show, do my show and deal with the public. As people began to wander through the show to get a break from the emotion of it all, they were running headlong into my images. Pennsylvanians began to make emotional connections with my artwork and TV pictures of evacuees. Viewers made me the point person for information, even though I had no clue what was happening at home.

The following week during my next show in Washington, DC, the same thing happened. One woman, an attorney, told me bluntly: "Go home right now and figure out a way to allow your artwork to tell Mississippi's story." It was a directive from the public to document the hurricane's effects and residents' recovery. Karole and I brainstormed the whole drive home. I thought about it. This is what I do every day: I document Mississippi through my artwork and introduce our community to the nation. I called my friend and congressman, Chip Pickering, to set a project in motion called "Backyards and Beyond: Mississippians and Their Stories". The idea was to establish a large exhibition of eighty paintings that will travel to museums and galleries for decades to come. Our goal is to raise awareness of what Mississippians have lost and to continue to encourage volunteerism and fundraising, not only for Katrina victims, but also for survivors of disasters to come. My paintings will be paired with live field recordings documenting the lives of Mississippians—black, white, Hispanic and Vietnamese—as they struggle to deal with life in the aftermath of Katrina. My images will be reflections of people dealing with their lives on a daily basis, showing the devastation either directly or indirectly and how we have adapted to our new life. It will be a challenging endeavor, personally and professionally.

When we arrived in our hometown of Jackson, we were overwhelmed by the damage in our city. One woman in my neighborhood was killed while reading in her bedroom, when a tree fell through her home. Jackson is 150 miles from the Gulf Coast. Seeing huge trees uprooted and splintering our neighborhood, blocking our streets, looking as if a bomb had exploded, led us to feelings of dread. We realized the pain on the Gulf Coast that we were seeing through the media was only scratching the surface.

Two weeks after Katrina landed, following crews still clearing the roadways, Karole and I began making our way from Jackson to the Gulf Coast. On Sunday, September 18, I began photographing for the project in Carriere, Mississippi, about fifty-five miles from the coast, at a Baptist church service. Their church building had been so badly damaged that they were meeting in their kitchen. Their skating rink, with its blown-out walls, had become a distribution center receiving trucks from around the nation. Piles of clothes, diapers, bottled water and canned food replaced third-grade skaters and Sunday family gatherings. That moment, that reality did not seem possible. But the evident hard work and presence of the people showed it was real. There was need out there beyond the collapsed walls of the church…beyond the morning service…beyond the off-key singing. The pastor upheld his responsibility to shepherd. The congregation was upholding theirs, to serve. As we would soon see, congregations from around the nation and service-oriented groups and individuals would flood our state with much-needed aid: physical, mental and spiritual.

Once on the coast we listened to story after story of loss and thankfulness. We encouraged the storm's victims with quiet hugs and the promise to help tell their story to the nation. A walk through a fairly new home revealed a swampy mess that was buckled and torn, sitting seventy-five feet from its original foundation. What we shared were the common dreams of every Mississippian. This was our space for all that was meaningful in our lives, our collections of all shapes and sizes: saddles, guns, toy fire trucks. These personal creations were made from store-bought ceramic molds, but now the molds are strewn in small stacks around the yard, waiting for attention. Attention that may have died during the flood…attention that will for many years be put into the rebuilding of their basic needs. There's no time for past "hobbies," only time to plan for a new life.

It has been said, "Nature is the arena in which reasonable men meet God". To be a human is to be on a journey, and to be a survivor of a hurricane is to advance in that journey. There are always dangers, difficulties and heartache to face along the way. There are also blessings, encouragements, comforts and friends to enjoy, too. But the point is that there is no going back. There should always be progress. We know God's grace and strength is always sufficient for us. Mississippians feel they are in a battle worth fighting, rebuilding their lives in a place that they love. We persevere in any struggle. We are nourished by our family, friends and neighbors. Recognizing our desperate state and our need for others' compassion is sometimes a difficult thing to accept. However, the thread of resiliency, spirituality and strength is always evident in the people of Mississippi.

There is something beyond the moment in the eyes of my fellow Mississippians. They just keep going through it all. Generations of proud families won't be run out by nature. We went through a place of fire and water, yet our hope at the end of it all is a place of abundance, because we have the ability to work hard and the faith to know we are not alone in our battle.

The Mississippi Gulf Coast is trying to survive. Many of the homes that dotted the region are now slabs—grey reminders that families once lived happily in the area. Other homes have seen scores of volunteer workers, top to bottom, cleaning and repairing the repairable. Many families have said their goodbyes and are waiting for the next step—waiting for the rest of their lives to begin.

As an artist I can't help but look through my camera lens and think, "This is history. I might as well be watching people fleeing the Dust Bowl. We are all living through history." As a Mississippian I can only hope I have encouraged you to consider helping rebuild our state. Come visit Mississippi and enjoy its people and treasures. We have found that news stories move on. Other things quickly command our nation's attention. But Mississippi's story goes on. We daily need people to find creative ways to help. I have a responsibility to tell the story people want told. The biggest challenge for you is to continue to listen and respond.

20

Nancy Burris Perret

Nancy notes that she writes this chapter from two perspectives. She gained a broad view of the aftermath of Katrina based on her experiences as a consultant for the Emergency Operations Center (EOC) for the City of New Orleans. A New Orleans resident who personally observed the response of government, military, nonprofits and for-profit entities to this unprecedented natural disaster in our country, Nancy also writes about the effects of Hurricanes Katrina and Rita on private citizens' lives. Using Abraham Maslow's Hierarchy of Needs as a model, Nancy examines the response of hurricane victims such as herself, as they move from struggling to meet basic needs (food, shelter, safety and so on) to recapturing greater control over their own lives.

—Dr. Rob

I am a self-employed consultant and trainer living in the New Orleans area; I help organizations with strategic and tactical planning and leadership and team development. In the year leading up to Katrina, one of my projects was to work with the New Orleans Health Department to develop a plan that would address the most pressing public health issues in the community. We were just finishing the final draft of the culminating document when Katrina hit the area on Monday, August 29, 2005.

I had evacuated on Friday, not realizing I was doing so. I had to work in central Louisiana on Saturday morning. By the time my work was finished, I would not have been able to re-enter the New Orleans area, since the government officials had already put contraflow into effect—an evacuation technique that redefines the flow of interstate traffic, forcing all lanes on both sides of the interstate to *only* go outbound. This concept, coupled with a staged evacuation of southeast Louisiana that allows the southernmost communities to evacuate first, was first used a few years before and had been adjusted to address some bottlenecks that had occurred in that evacuation.

A seldom-articulated fact outside of Louisiana is that the evacuation preceding Katrina was the most successful mass evacuation of an American city in history—with no appreciable gridlock or loss of life. Widely issued criticism about the delay in ordering a mandatory evacuation of the City of New Orleans proper must be tempered with the phased evacuation plans that all the area parishes had agreed on, influencing the first moment that a mandatory evacuation could be issued. Additionally, a voluntary evacuation had been widely recommended as early as Saturday morning, when the track of Katrina had changed from focusing on Florida's panhandle to southeastern Louisiana and Mississippi.

Mass evacuations can be a tricky thing, as evidenced by the disastrous evacuation of Houston a month later, when Hurricane Rita briefly threatened that city. Lack of clear planning and communication to the general public about how to safely evacuate were evidenced in Houston, but not in southeast Louisiana. The sole defect in this evacuation (albeit a large single failure) was having an adequate plan for transporting and sheltering the poorest residents of the area who had no transportation of their own.

To get a quick sense of the scope of this disaster, imagine evacuating one-quarter of your state's population. Next, imagine great damage to eighty percent of the housing stock that sheltered that population, and the complete destruction of the infrastructure supporting that population.

I stayed in central Louisiana, and was joined by my then-fiancé, his sister and his son. We watched the crisis in New Orleans unfold following Katrina's landfall—a crisis initially created by the hurricane but made infinitely worse by the failure of the levee system designed to protect the city—and feared the worst about each of our homes. It would be over a month before we could confirm the extent of damage to each of our homes. My home in Metairie sustained limited damage. My then-fiancé's home in Slidell was inundated with six feet of water and was hit by four large trees. It was later demolished. His apartment in Gulfport was literally gone, with no evidence that he had ever been there. His office sustained twenty-two feet of water and was uninhabitable for nearly nine months. His son's home was relatively unharmed, although his son was denied access for nearly a month as the city recovered. His sister's home was inundated by eight feet of water that remained in the house for three weeks.

I had the amazing opportunity to staff the Health and Human Services desk of the city's Emergency Operations Center (OEC) beginning in mid-September, when the New Orleans area was only beginning to open up. Walking in the door of this facility three weeks after Katrina hit, I found a well-organized room designed to facilitate the rebuilding of the city. Present in the room were military planners and logisticians, a volunteer group from the state of Virginia, representatives from every utility company (electric, gas, water, sewage, and telephone), government representatives for each critical element of the city's infrastructure (education, environmental safety, parks and parkways, debris removal, public housing, *et cetera*), certifying agencies for private industry (such as healthcare, food establishments and pharmacies) and key nonprofit organizations that ultimately would coordinate the response of hundreds of local and outside nonprofits.

In spite of the city's general disarray, this center was able to make data-driven decisions as a result of an amazing level of collaboration, as people shared information that proved critical to the emergency response efforts.

The presence of the military and the volunteer team from Northern Virginia were essential to the operations of the center at this time. It is safe to say that *every* citizen of the area was immediately and profoundly impacted by the results of Katrina. While New Orleans is well known as a tourist destination and a city of great diversity, few realize that most of the people in the area are connected through family lines for many generations. As a result, it became immediately clear that, if a person's home and immediate family was safe and secure, one didn't have to look far from that immediate family to find someone who had lost their life, their home and/or their employment. The impact of the nearness of the trauma would reveal itself throughout the emergency response, and continues to reveal itself during recovery.

It became quickly evident that we *all* were impaired at some level. Regardless of how well each person was coping with the continually unfolding drama, we were clearly dependent, as individuals and as a community, on the outsiders who were there. The Emergency Operations Center structure and the presence of the military and the Northern Virginia team provided us with a command and control system that was essential for smooth operations and timely decisions. Their goal was to stay in town only as long as we needed them, although *our* thoughts about how long that was and their leadership's ideas on the subject often differed widely.

It was clear that all of the necessary leaders and experts needed to communicate, so that they could make informed joint decisions. Many of the positional leaders within the community (governmental, nonprofit and corporate entities) were not in the city during and immediately following the disaster, and were directing actions from afar. Providing them with the most up-to-date information, which was changing from minute to minute, was essential to helping them make informed decisions in a timely way. Daily briefings gave all of us in the room information about all the other elements of operations, and made it very clear how connected we were in restoring normalcy to the city.

One of the tragedies during this period was that, as the existing infrastructure returned to the city, many did not appreciate the value of the EOC and resumed operations at their primary (in many cases, temporary) facilities. While this was essential to the restoration of the city, the value of the EOC diminished as decision-making moved back to the primary operations sites. Because the leadership did not understand or appreciate the value of this centralized operation, it did not use the EOC as well as it might have.

As people returned to the city—slowly at first, and then in large numbers—you could tell at first glance how long they had been back. The magnitude of the devastation in the community is incredibly difficult to communicate to those who have not been here. Ninety thousand square miles (roughly the size of Great Britain) were devastated by Hurricane Katrina. Then, less than a month later, Hurricane Rita increased the devastation in southern Louisiana, laying waste to a significant area of southwest Louisiana and re-opening minimally repaired levees in the New Orleans area.

When one first returns, the impact is devastating. The level of services that we Americans have grown to expect remain absent. We truly are re-inventing southern Louisiana and the Gulf Coast of Mississippi, from the ground up. This is very difficult for people to understand; in their first few weeks (and sometimes months) they often appear shell-shocked, as they begin to grasp how much their lives have been altered by this series of disasters. People are returned to their most basic level of functioning very quickly. Survival becomes the primary focus. Next, the focus is safety and shelter.

Still, people in the area continue to speak of disturbed sleep, a very high level of depression, an inability to get the resources they need to rebuild their lives, and a lack of control over their lives. These elements have influenced our ability as a community to move forward.

A feeling of helplessness and inadequacy often emerges as people who are accustomed to making their own decisions and managing their own lives find themselves unable to do so. *Self-determination* is an assumed state for most healthy, productive adults. But government inaction at all levels, and the slowness of response of insurance companies, have together made this impossible for many residents of the area. Dependence on others for things that we have been used to handling on our own is debilitating in so many ways. Finding ways to re-empower people as quickly as possible (within a clear command and control structure) is essential to rapid restoration of a community.

Lack of *clear direction* from political leaders at all levels, and embarrassingly slow distribution of allocated funds to help in the rebuilding process, have exacerbated the situation. It is essential that leaders at all levels clearly articulate the situation, including saying things that may be unpopular. Even if the news is something we don't want to hear, it's better to hear it than to remain ignorant. (A clear example is the oft-stated potential solution of "shrinking the footprint of New Orleans". Not a single political leader has articulated the fact that this simply means reducing the area that will receive city-level services. Parish-level government will still be available to provide a lesser level of services, yet this has not been made clear to people. The public has perceived this concept as a declaration of where people will be *allowed* to build and live.)

Positional leaders were not the primary leaders during the immediate aftermath of the storm, except in the case of the military. Leadership emerged in some very unlikely places, and shone brightly during our immediate response. Clear communications, the ability to make decisions in a rapidly changing environment with limited constant conditions, and the ability to remain calm and focused were essential. Future crisis response efforts would benefit from a method of *testing positional leaders to assess their ability to lead.* Too many positional leaders attempted to lead when they were severely impaired—a circumstance that proved disastrous, in some cases.

The entire community was reduced to operating in survival mode for many months, and in some ways, it remains in this state. Loss of nearly two hundred thousand homes in the areas affected by Hurricanes Katrina and Rita has staggered every system designed to aid people in the aftermath of catastrophes. *Two hundred thousand homes!* In addition, the small business community, which represents about ninety percent of the businesses in the area, has been devastated with loss of facility, customer base, and resultant revenue streams.

Decisive action and rapid, direct response to the people who need it, was woefully missing—and remains so. While billions of dollars of federal assistance has been allocated, and in some cases distributed, little of this money is in the hands of the people, nearly a year after the disaster. The response of the local and national nonprofit community has been the greatest aid to recovery since the departure of the military.

The functions of the Emergency Operations Center were critical to the early recovery of the city—and, in my opinion, would have had greater benefit to the city's long-term recovery if the leadership had completely understood the way the EOC operated and perceived its value.

Individuals within the city are reclaiming their lives—and their ability to self-determine their future—as they wrestle through the many aspects of rebuilding. As individuals restore their lives, one bit at a time, you can see in their faces their increased ability to cope with the continuing difficulties of living in a stricken area. Moving through the first of Maslow's stages (first survival, then safety and shelter) is critical to this progress. Once people have found a shelter, even if it is a temporary shelter, they can begin to restore connections to family and friends, focus on their own self-esteem and return to a state of self-determination. Those who remain dependent on others to resolve their needs for shelter (whether that is FEMA, insurance, or public housing), remain "stuck" at this level—unable to continue the process of rebuilding. This progress, person-by-person, is essential to the community being restored. As long as we have inadequate or unaffordable housing, we will remain "stuck" as a community. Still, individuals are finding ways to move forward—with or without government assistance.

The resiliency and strength of the people who have re-settled in the area is remarkable. Their strength of character and courage will lead us forward as others return. There are incredible stories of courage, sacrifice and aid to others that continue to emerge.

The first visible "heroes" of Katrina were the emergency responders. Their actions saved thousands of lives. Additional heroes are visible every day in the community: the healthcare workers who have been stretched beyond human limits (yet continue to deliver aid to the wounded and infirm); the nonprofit aid workers who have volunteered their time and human spirit to help us to rebuild; the everyday citizens who have reached out to help neighbors in need.

Katrina has changed our lives as individuals, and as a community. For example, in the aftermath of Katrina, all of my clients were absent from the city. Complete loss of income stared me in the face. Through a series of events, my role in the Emergency Operations Center filled this gap of productive work. In addition, it provided me with an unparalleled opportunity to meet a team of professionals who knew how to come to the aid of a devastated people and community.

The relationships I've built and the lessons that I've learned will affect me for the remainder of my life. While I've always recognized the need for clear direction and strong leadership, I have seen people operating in the worst of conditions—and I've gained a lifetime of lessons about successful and failed leadership that will inspire me. In turn, I hope to inspire others through my words and actions.

Now, as I work to rebuild my life, I am far more aware of my own limitations. Decisions are more difficult for me. Managing my time has become more difficult. There are so many interconnected parts of my life that remain unresolved, and too many elements that remain dependent on the actions of others. I believe I'm ready to return to managing my own life, but I remain stuck in a couple of key areas that I cannot control, or even influence. Keeping my frustrations in check is a challenge.

I am also painfully aware that I flinch at the thought of another hurricane heading in this direction. I will follow the hurricane plans I have always obeyed, and if need be, I will evacuate well in advance of the storm. But I am far less confident that that evacuation will be short and that I will be able to handle whatever I find when I return.

Recognizing where people and organizations are, relative to Maslow's Hierarchy of Needs, can be helpful in aiding them in their own recovery. Helping others to quickly regain their prior ability to manage their own lives is the greatest help possible. We need not make disaster victims dependent on others throughout their adult lives. Giving people an active role in restoring their lives, regardless of the size of the disaster, is essential not only to recovery, but to their ability to function at their highest possible levels.

This approach can allow people to take ownership of their own future whenever they are able. This doesn't mean that we don't need government assistance, but there is a difference between assistance and dependence. For those who *are* dependent on government programs to survive, this need will continue after a crisis. But creating a new group of government dependents is neither productive nor healthy.

Building on people's natural resiliency and strength—even when that resiliency and strength is temporarily impaired—will speed recovery for both individuals and their communities.

21

Evangeline Franklin

While coastal Mississippi offered healthcare primarily in the form of private practice and a few health clinics, the situation in New Orleans was very different. One of the greatest successes in the Mississippi healthcare system was the cooperation of competing pharmacies and insurance companies as they helped patients recreate their medication records, using computerized billing systems. In New Orleans, Katrina destroyed so many records and facilities that it was a great challenge reassemble medical records of the chronically ill. It is clear that many basic healthcare system needs can be covered through technology. Below you will find an insider's perspective with brief history of the health care system in New Orleans, how it reacted with innovation and courage, and ideas about where it is going.

—Dr. Rob

Pre-Katrina New Orleans: A Community Already at Risk

Before Katrina, conditions in the New Orleans healthcare system were already dire; Louisiana ranked as one of the five worst states for infant mortality, premature deaths, and smoking and cancer fatalities. Katrina devastated the city's healthcare infrastructure, depleting supplies, destroying all types of paper and computer records, crumbling facilities, compounding health problems, creating unsanitary conditions and countless injuries and evacuating staff. It was a nightmare. For the poor, the uneducated, the homeless, the immigrant, the uninsured and others who were disenfranchised, access to quality healthcare was even more difficult.

While disparities in health services have long been a problem in New Orleans, the situation has worsened in recent years. Pre-Katrina New Orleans was a majority black city with extreme levels of poverty, resulting in a de facto caste system of healthcare—a situation that was, of course, not unique to that city or state. In addition, beginning in 2004, state cutbacks in funding for the only state charity hospital system in the nation led to further dramatic reductions in ambulatory care services for the poorest residents. The Walk-In Ambulatory Care Clinic W-16 at the Medical Center of Louisiana at New Orleans, formerly known as Charity Hospital, (which had, for generations, been appropriately used for non-emergency medical problems and available to uninsured and insured patients, alike) and the emergency room at Bywater Hospital, formerly St. Claude General Hospital, were closed, further reducing access to non-emergency care for the city's uninsured and underinsured by more than two-thirds. These facilities handled nearly 56,000 individual doctor visits annually, mostly sourced from zip codes 70117 and 70119 (the neighborhoods that contained the city's poorest people). With the removal of these sources of healthcare, those in need of free care began to show up where they could find the easiest remaining source of care on demand and without charge, and so emergency rooms became increasingly congested. It became a huge challenge for the system to help patients transition to stable alternative sources of primary and specialty care, and to pay for that care, once found. This situation has been disruptive, to say the least, and has created a significant challenge for the Charity Emergency Room system, as well as for other emergency rooms in New Orleans.

Poor health status in pre-Katrina New Orleans and elsewhere in Louisiana reflects more than just an inability to access the healthcare system. It is also the result of individual and community factors, such as ingrained behaviors regarding eating and exercise, language barriers, levels of education, community crime rates, employment status/income level, cultural and regional traditions, family values and attitudes, environmental and political factors—all of which contribute to poor overall health and worsening health disparities. In the case of New Orleans, many of the city's critical assets are also its tragic flaws. After all, in New Orleans we have a popular saying found—and practiced—everywhere: *"Laissez les bontemps rouler!"* (*"Let the good times roll!"*)

Unfortunately, the city's fabled cuisine features high-calorie, high-cholesterol, refined sugar dishes (big breakfasts with eggs, grits, bacon and biscuits; fried shrimp, fried oysters, fried catfish; hot sausage "po'boys" slathered with mayonnaise; pecan pralines made with pecans, butter, sugar, and cream; and bread pudding (made with leftover French bread, eggs, cream, sugar, and spices, and topped with a rum or whiskey sauce). In addition, policy to improve the nutritional status of those who could not buy food was implemented in the 1960s. A landmark book written by Robert Coles, *Still Hungry in America*, chronicled the lives of citizens in rural areas of the Gulf States, where low levels of poverty resulted in poor nutritional status, and in some cases, near starvation. Various supplemental food programs were implemented to improve the health of the poor, but the pendulum for those communities has now swung the other way. Louisiana, and the South in general, now lead the nation in obesity and the health problems connected with that condition: diabetes, hypertension, and cardiovascular diseases.

And this culture's health problems reach beyond those caused by the attractive cuisine. A relaxed environment, and the relaxed behaviors it produces, can lead to substance abuse, poor judgment in sexual activity, and drinking and smoking habits, to name a few. In the case of unprotected sex, serious communicable medical problems like HIV are taken back home, in and out of the city and state.

The city's employment structure is a factor as well. Essentially, New Orleans is a port city with an economy also based on tourism. The level of education required for these jobs and the benefits attached to them are comparatively low, and generally, employees do not understand how to use health benefits most beneficially. Additionally, New Orleans' historical lack of a manufacturing sector has prevented the development of a strong labor movement, with its demands for healthcare.

Thus, even prior to Katrina, a grim picture of the health status of a poor, predominantly African American city emerges, demonstrating the challenge of breaking a multi-generational cycle through aggressively engaging in a plan that targets the disparities of health status by race and class.

Post-Katrina New Orleans: Damage to the Health Infrastructure

Following Katrina, over seventy-five percent of the city was affected by flood water that had become brackish from ocean surge and that was contaminated with sewage, dead bodies, polluted household chemicals, lead-based paint and vehicular fluids. In the heat of the summer, that water could not naturally recede for three weeks.

In that flood plain were the homes and offices of many healthcare providers. The healthcare delivery system—in the form of hospitals, doctors' offices, nursing homes, labs, x-ray facilities, treatment facilities, behavioral health facilities, supplies, vehicles, medical records, pharmacies, equipment, supplies, drugs and medical records—virtually vanished for nearly all New Orleanians, public and private, white and black, rich and poor.

Most of New Orleans' hospitals are located in the center of the city and thus were damaged or flooded as a result of the disaster. Over half of all hospital beds in Louisiana and one of two Level One trauma centers were lost, due to their location in New Orleans. By November, three months after Katrina, only three hospitals were fully functioning in the greater New Orleans area, and none were located in the City of New Orleans. Only one New Orleans hospital, Touro Hospital, reopened its doors soon after the storm, and the only care available at that facility was emergency room treatment. Three hospitals in adjacent Jefferson Parish were operational, but their ERs were overwhelmed with patients who had stayed in the Parish or relocated there from devastated areas.

After Katrina, this limited care was free of charge. Some hospital beds were available through the Combat Support Hospital (successor to Mobile Army Surgical Hospitals, or MASH) and the *USS Comfort*, a floating hospital supported by the Navy. Only two of the nine New Orleans Health Department clinic sites were operational. The clinics of Excelth, Inc. and Daughters of Charity were all destroyed. The medical records of all of these clinics were destroyed by flooding. Most pharmacies in New Orleans were also destroyed, and government shipments were delayed. Numerous nursing homes and assisted and independent living facilities were also located in the flooded area, placing their population at risk because of the lack of utilities, sewage, and water. Ambulatory healthcare facilities were also badly depleted in the aftermath of Katrina, with many facilities sustaining damage sufficient to render the facilities unusable for medical purposes. Because of this, a substantial number of healthcare workers had relocated after the storm and could not return because there was no place for them to live and work. Moreover, what patients there were often could not pay for their care. It is reasonable to anticipate that the vast majority of healthcare workers facing this situation may choose not to return for some time, or possibly ever.

The Diaspora

With most New Orleanians' homes completely devastated from floodwaters, most residents were unable to return after evacuation. They had been separated from their family and pets, and were unable to get clear and reliable information about the state of affairs in the city of New Orleans for many months.

This resulted in a wide distribution of persons throughout the country, or a diaspora, of massive numbers. Under the civil authority rules, people—including those who had evacuated to locations outside of New Orleans before the storm—could not return. Those who had remained in New Orleans, and were housed at the Superdome or the Convention Center, could not stay for any length of time. They were transferred by bus to Dallas, Houston and San Antonio. From there, many dispersed even further.

As an evacuee to Dallas, I personally learned what had to be done to assist and support evacuees. Many communities had to rapidly establish shelters that included emergency medical and social service operations for the large influx of New Orleanians, regardless of their arrival time in these cities. Patients needed to refill medications and find new doctors to care for their chronic illnesses. They also had to deal with mental trauma, and find new emergency rooms and hospitals when they suddenly became medically or emotionally unstable. Those living in a shelter had to locate a new place to live and find new social services and new schools for their children. They also needed to look for new employment. Medical facilities, mental health facilities, social services and providers of all types rapidly became overwhelmed. As shelters became overcrowded, patients frequently were transported, without their consent, to other shelters. During this process, families were separated, increasing the stress of the situation.

It became clear in the days following this mass evacuation that the vast majority of the citizenry seeking medicines, medical care, and behavioral health interventions for chronic and acute problems could no longer depend on having access to the information that resided in their doctors' offices or at the hospitals. Patients were completely displaced from their physicians and other healthcare providers, and medical professions strained to treat them, for a variety of reasons.

Of the evacuated patients who needed some kind of medical and/or behavioral healthcare, many only had a general idea of their diagnoses, and they did not understand the medications and therapies they needed to remain medically stable. To complicate matters further, it was difficult for medical professionals to get this information. For several weeks there was no way for healthcare providers, or anyone, to call phone numbers in the 504 area code. Even when phone lines to that area were opened, there was often no one left in New Orleans to pull a file (a file that had likely been destroyed, in any case, as most paper medical records were either under water, or had become wet and moldy in the humid environment). There was no one in oncology to call to get cell counts, reactions to chemotherapy, and the monitoring studies that go along with the treatment. There were no available surgical notes describing the anatomy of a procedure, no notes indicating the total number of radiation treatments given to a lymphoma, and no x-ray films to review. There was no one to call at the academic medical center transplant program to find out the correct name, schedule, or source of the experimental transplant drug for the renal transplant patient evacuated from his home via the Superdome. New Orleanians scattered throughout the U.S.—who were not medical professionals—were required to reconstruct their medical histories from memory. And the professionals who previously cared for these patients had lost their homes, offices, jobs, records and the hospitals they practiced in.

The Immediate Help

During the post-evacuation period, New Orleans was occupied by a variety of military personnel who valiantly provided a variety of services, including healthcare. The few doctors who remained in the city worked with the military medical personnel to make certain that minimum care was being provided to working military personnel and citizens left behind in less-damaged parts of the city.

Aside from immediate and long-term mold and the lack of water, food, shelter, and sanitation facilities, there was concern that the prolonged flooding might lead to an outbreak of health problems among the remaining population. In addition to dehydration and food poisoning, there was the potential for communicable disease outbreaks of diarrhea and respiratory illness, all related to the growing contamination of food and drinking water supplies in the area. After dewatering and house-to-house searches for dead bodies and animals, public health concerns centered on acute environmental hazards related to houses standing in water for several weeks (mold, bacteria, concealed rodents, snakes and alligators). This was combined with ruptured sewage lines, refuse, structural instability, debris, the lack of sanitary water (what water they had was usable only for flushing toilets), as well as a lack of gas and electricity. As the dewatering of the city proceeded, it became evident that a massive assessment of all infrastructural components was required to protect the health and safety of the citizenry. Martial law was imposed, and citizens were prevented from returning. Before the hurricane, government health officials had prepared to respond, and the Centers for Disease Control and Prevention (CDC) began to send medical emergency supplies to locations near the worst-hit area within forty-eight hours after landfall. The New Orleans Health Department directed the state and CDC personnel in developing a surveillance tool to help keep track of potential outbreaks.

National Guard, military and U.S. Public Health Service personnel convened regular meetings of the available medical provider leadership to assess the state of physical and human assets. The availability of vaccines and emergency services was considered a key first accomplishment. Few local services were able to open, and care was augmented by the military and out-of-state volunteers permitted to provide care under an executive order by Governor Blanco.

Supplies shipped by the CDC's Strategic National Stockpile (SNS) and other private relief organizations provided pharmaceuticals. The CDC's supplies served an estimated thirty acute care hospitals south of Interstate Highway 10, and volunteers organized around its "contingency stations" to become temporary stand-ins for hospitals, warehouses and distribution facilities damaged by the storm.

Within days after landfall, medical authorities established contingency treatment facilities for over ten thousand people, and plans to treat thousands more were developing. In some cases, partnerships with commercial medical suppliers, shipping companies and support services companies ensured that evolving medical needs could be met within days, or even hours.

There was concern that the chemical plants and refineries in the area could have released pollutants into the floodwaters. People who suffer from allergies or chronic respiratory disorders such as asthma were susceptible to health complications due to toxic mold and airborne irritants, leading to what some health officials have dubbed "Katrina Cough." On September 6 it was reported that Escherichia coli (E. coli) had been detected at unsafe levels in the waters that flooded the city. The CDC reported on September 7 that five people had died of bacterial infection from drinking water contaminated with Vibrio vulnificus, a bacterium from the Gulf of Mexico. Wide outbreaks of severe infectious diseases such as cholera and dysentery were not considered likely because such illnesses are not endemic in the United States.

Health Recovery Week at the Audubon Zoo: February 6-12, 2006

In early November 2005 I was assigned by Dr. Kevin Stephens, director of the New Orleans Health Department, to be project manager of an ambitious outdoor primary care center that was eventually incorporated into a program called Health Recovery Week. The goal was partly to empower patients with their own health information in the case of future evacuation. The health information collected then could be loaded onto a portable digital device so that patients could be directed to follow-up care. Kyle Park, communications specialist and chief of logistics for the Strategic National Stockpile Delivery in the Dallas/Fort Worth Regions of the Texas Department of Social and Health Services, was brought in to blend the clinical side with the IT side of the project. This collaborative opportunity was supported by the Morehouse Medical College Hurricane Recovery Center, Remote Area Medical Volunteers, Intel Corporation, the American Academy of Family Physicians, Solventus, Inc., and local medical practitioners. It would provide healthcare and behavioral health and social service information on a first-come/first-serve basis. At the end of the event, patients would receive an electronic record that they could give to any physician, local or remote.

It was more than a simple health screening. Participants were offered full office services, including lab tests, Pap tests, and follow-up appointments. Patients were quickly given critical and abnormal test results so treatment could be sought. Mental health providers circulated throughout the crowd to help identify those with special behavioral needs and to guide patients to the appropriate locations. Volunteers participated from local colleges and universities and local safety-net clinics. Many of the practitioners—most of whom were from out of state—said they had never seen the level of disease present in the patients coming to the event. Health Recovery Week also documented the vast scope of need in the city and surrounding areas.

Over five thousand citizens—some from as far away as Baton Rouge and Mississippi—attended the event and were provided dental, medical and optical care, including free glasses and thirty days of whatever prescriptions they needed (over four thousand prescriptions were dispensed). On leaving, they had available to them a CD-ROM or USB memory drive that stored their health information in a portable, digital format. (The event revealed that patients wanted their health information in this format.) All clinical and non-clinical staff of the New Orleans Health Department participated in the event and was trained on the software. Many patients became volunteers.

There were many requests to repeat this event before June 1, the opening date of our annual hurricane season, so that patients could receive their portable health information before being evacuated. I branded a new saying for the City of New Orleans for continued use by the Health Department: *"Laissez la bonne santé rouler!"* ("*Let the good health roll!*")

The Rebuilding

New Orleans is faced with a repopulation as former residents return and new residents arrive. No doubt there are physicians among those inhabitants of the city applying for permits to rebuild their homes and offices. But physicians need hospitals, support facilities, labs, x-ray units, and pharmacies to provide any level of medical care. These facilities will likely return, but not soon. Layoffs had occurred in the City of New Orleans workforce (Health Department staff for its clinics went from two hundred and fifty to seventy-two), as well as in the school system (of course, this resulted in large numbers of persons losing their insurance). Monies for various health programs were shifted out of the city at the same time the rate of persons returning was increasing. Despite the increase in the number of public clinics open in all of Region One (Orleans, Jefferson, St. Bernard and Plaquemines parishes), an analysis of office and hospital locations in flooded areas confirmed by Internet registries indicated that only fifteen to twenty percent of private physicians had returned to practice by early spring. Many patients did not know where their doctors were and were not looking for replacements of those valuable relationships. There were virtually no dentists practicing at all. Two organizations are currently allowed to funnel in volunteer medical staff, clinical assets, and pharmaceutical donations for the distribution of medication to those without insurance.

The New Safety Net

Despite our recognition of the health disparities between races, genders and class structures, disasters do not make these distinctions. It is how we prepare for and respond to those disasters that will express our dedication to the equal treatment of all people. The extensive disruptions in the New Orleans economy, altered living conditions, and the desire and ability of individuals to live in the city will have a great impact on the future demographics of New Orleans. The city is already filled with a mix of strangers—many non-English-speaking—and newly unemployed neighbors, while some neighborhoods have permanently lost families because opportunities seemed better elsewhere. Exact changes cannot be predicted at this writing and would be speculation at best. But with the likelihood that a vast proportion of physicians will not be returning to a city that has less and less cash and insurance to pay for healthcare, there are concerns about reconstructing a safety net that must care for a large—albeit different—mix of patients.

It is reasonable to assume that many of the public health concerns documented in the August 2005 unpublished report I co-authored for the New Orleans Health Department, "Getting People Healthy in New Orleans (Franklin, Hall and Burns 2005)," will continue to exist post-Katrina, although perhaps in different geographic clusters and rates. The citizens returning to New Orleans and its new citizens, who will come to this "frontier town" with new kinds of employment opportunities, will require a healthcare delivery system that addresses probably the same medical problems but with culturally competent service. We must also remember, and try to address, the longstanding health problems that have plagued our traditional population.

Personal Empowerment and Health Technology

American medical culture is characterized by resistance to a truly patient-centric model. Standard organized medical practice for all patients—from small and large groups, urban and rural, private and public safety-net providers—is largely driven by the traditional physician-centric model based on paper medical records housed in the provider's office (on shelves which usually stand on the floor). Even a small amount of standing water lasting for many weeks in hot weather will ruin charts on the shelves. Paper medical records could be lost forever by flood, fire or other forms of destruction. Personal ownership of a portable health record allows patients to view records themselves, and to have them available twenty-four hours a day for visits to other doctors—by appointment, when arriving unconscious at an emergency room, when traveling, or during evacuation.

The key elements of a portable health record include:

- Optional password protection to avoid unauthorized access to the device
- Appropriate disclaimers and instructions for use
- Disease and status of stability, as documented by certain lab tests
- Medications/therapy for those diseases and compliance patterns
- Previous electrocardiogram and radiologic image (x-ray, echocardiogram, ultrasound, and heart catheterization)
- Implanted devices, transplanted organs, etc.
- Names, email addresses and phone numbers of treating physicians

- Names, email addresses and phone numbers of experimental drug or chemotherapy vendors
- Next-of-kin and emergency contacts
- Dates of when the information was entered and revised

This disaster provided an opportunity to make sweeping change toward the improvement and preservation of public health via inexpensive information technology models and, as a secondary outcome, to decrease the medical errors that have been documented in many studies. There is agreement among health leaders in the greater New Orleans area that this information technology is critical and will help greatly in subsequent hurricane seasons.

Conclusion

There are some very disturbing weaknesses in our healthcare system that Katrina has underscored. However, the hurricane also exposed (or re-exposed) those weaknesses as opportunities. Katrina provided us with a worst-case scenario to see how fragmented and unprepared our entire healthcare delivery system is when disaster strikes. In addition, Katrina also demonstrated both our medical system's overall inefficiency during normal times, and how fragile the communication tether is between healthcare providers and their patients regarding essential information. A vast number of patients, emotionally paralyzed by the loss of their homes, family members, pets, and belongings, were sent to distant, unknown cities. Many were unprepared to obtain the help and care that they needed. Many returned to New Orleans similarly unprepared.

Hurricane season is an annual event in New Orleans. In many parts of the country other threats are present, such as earthquakes, volcanoes, floods from rain and melting snow, and blizzards. In addition, terrorist threats like 9/11 are real possibilities. What we need to learn before the next disaster strikes in our communities and elsewhere can be summarized as follows:

- Mass evacuations should follow certain procedural standards created prior to this disaster. These standards should predict, as well as possible, where relocation will occur, so as to prepare temporary shelter facilities and appropriate medical services in those locales.
- As much as possible, provide current and accurate information to those who are impacted by the disaster.

- Take into full account mental and physical health problems that disasters and forced displacement produce.
- Health recovery events, such as those organized in New Orleans, are effective short-term stopgaps.
- Deliver portable health records to patients, and encourage physicians to institute compatible electronic medical records systems.
- In order to restore post-disaster medical facilities, establish systems that will assure the "safety net," create inducements for medical workers to return, and ensure the replacement of support services.

This chapter was written in May 2006, a month prior to the onset of New Orleans' annual hurricane season. We should know soon what we have changed and how we will do the next time.

Sources:

Connolly, Ceci, 2005a. "N.O Health Care: Another Katrina Casualty". *Washington Post.* Nov. 25.

Barringer, Felicity. 2006. "Long After Storm, Shortages Overwhelm N.O.'s Few Hospitals". *New York Times.* Jan. 23.

Nossiter, Adam. 2005. "Hurricane Takes a Further Toll: Suicides Up". *New York Times.* Dec. 27.

Connolly, Ceci. 2005b. "Katrina's Emotional Damage Lingers". *Washington Post.* Dec. 7

22

Summary and Recommendations

What do these stories have in common? How did these people emerge from a disaster with the conviction that there can be success in the face of adversity and hope in the presence of overwhelming loss? What motivated them to take initiative, where some lost faith? Vangy and I are marvel at what went right in the busy days and months that followed Katrina. We are proud to have worked with these authors, and with countless others whom we met during the chaotic aftermath and recovery phase. It is our sincere hope that no one must go through the trauma of a disaster, but we also hope that we can motivate the reader to get prepared. The following chapter will help you plan, so you can prevent, prepare for, respond to, and recover from emergencies in your life.

—Dr. Rob

As this book was being written, the country was anticipating the one-year anniversary of Katrina. As we reflected on lessons learned from that disaster, we realized that we not only faced risk of another damaging storm; we also faced risk of crimes, war, terrorism and natural disasters.

As the five-year anniversary of the attacks of September 11 2006 grew near, we remembered the volunteer workers who risked their lives to save others, and the health difficulties they had suffered. We watched movies about the World Trade Center and Flight 93. We learned about a close call in August 2006, when suicide terrorists were apprehended in Britain after planning to explode multiple airplanes with improvised explosives concealed in sports drink, shampoo and baby formula containers. Our President stated that the country was at war with a "fascist Islam movement". The Center for Disease Control and Prevention worried about how to reduce the spread of a potential deadly virus, H5N1, which, we learned, could be source of a deadly flu pandemic.

Fortunately, all hazards share basic principles of preparedness, response and recovery. As we face the future, let's look at common practices that you, your family, your workplace and your community can put in place to ensure the safety of all involved.

Some very simple things can make a difference in your experience with a disaster. People who successfully meet adversity rise to the occasion, taking action while others waited for help. They had listened to advice from authorities. They had planned and prepared. When met with unforeseen difficulties, they improvised. They took care of themselves first, and then reached out to help others. They also accepted help, and they continued to work with relief agencies to improve their situations.

A fine example of success in the midst of disaster is the well-run aid effort by volunteer-staffed organizations that traveled to the Gulf Coast to help Katrina victims. For years, faith-based organizations have assisted in the aftermath of local disasters, and have taken distant mission trips. They spontaneously filled the gaps in an almost miraculous way. We watched the Amish and Mennonites use their barn-raising experience to rapidly repair roofs and raise new houses, and we admired their hard work as they moved debris. The Lutherans built tent cities and infrastructure support for other volunteers. The Methodists offered counseling. The Remote Area Medical group assisted with medical missionary activities. And, from across the nation, countless other faith-based and secular volunteer groups pitched in, as well.

This included a group from the Mayo Clinic, which staffed temporary medical primary care clinics in rural Louisiana. Domestic and foreign companies, large and small, donated items for use. While the government and the Red Cross continue to help meet the immediate needs, it is the spirit of a generous and informed faith-based and business community that helps backfill where the need is large.

These successful organizations are staffed by members of the Freedom Generation. They are citizens who have taken the initiative to help themselves and others, because they know that they can make a difference. Those in the Freedom Generation include Scout leaders and Scouts. They become EMTs or CPR instructors or students. They join these organizations or committees at work.

They prepare their families and encourage their community leaders to spend the resources to take the lead.

We challenge you to become a Freedom Generation member and improve your chances of writing your own "successes achieved" essay after the next man-made or natural disaster.

How do you begin? First, learn from the past. The recovery and aftermath of Katrina went poorly in so many ways, but it is important to remember that this disaster was overwhelming, and reached beyond the scope of anything local officials had ever experienced. This problem was compounded by a lack of preparation and planning among individuals, fueled by an erroneous belief that the government can take care of all citizens in any situation. It came as a shock, then, when citizens called 911 and found that the phone did not work, or that no one was there to answer the line. It was incomprehensible to them when they went to the grocery store and found that it wasn't there, or when they visited a clinic and discovered that the doctor had evacuated, or that the entire facility was underwater and their medical records were gone forever. Citizens need to increase their chances of future success by learning how to take care of themselves in any situation. They must learn from the mistakes and successes of the past, and plan accordingly.

If an emergency affects you, you must be the first to respond. If it affects those close to you, you must help them, as well.

The First Step: Make a Plan

Organizations and individuals face different challenges, but they each can benefit from this simple rule: Major successes result when plans are made and rehearsed. Planning with others creates channels of communication, familiarity and trust that become vital in an emergency situation, when unexpected circumstances arise.

Think of alternative strategies and be prepared to change course if your initial plan fails. Simplify your strategy and try to think globally, since you want to create a plan that will apply to a variety of disaster scenarios (professionals call this an "all hazards" approach). You can modify the plan to suit the specifics of the situation, as opportunity arises.

Prevention and Mitigation

Assess your risks. What is likely to happen in your region, neighborhood or home in the case of a volcano, earthquake or flood? Is your neighborhood at risk of gang violence, chemical spills or traffic problems? Is your home at risk of fire or flood? In your present location, are you likely to fall out of a window? Do you regularly travel by mass transit, and if so, what would you do in case of an en route emergency? Figure out the dangers that are most likely to occur. Ascertain which risks are in your control. Then, look for ways to contain those risks so that the likelihood of damage is reduced. You may not be able to stop a hurricane, for example, but you can plan to leave before it hurts you. You may not be able to get rid of violence on the streets, but if you must travel after dark, you can be sure to walk with someone else, and on well-lit streets.

Remember that simple prevention items, such as smoke detectors with good batteries, are as important as planning for the next big hurricane. Be sure to have these items on hand (and to replace batteries regularly, when applicable).

Keep vigilant. Look for the suspicious package and the out-of-place trench coat on a hot day. Know where the exits are in a movie theater or other public place. It is empowering to have an awareness of your surroundings and a contingency plan. You can be aware without being afraid.

Preparation

Start with yourself. Learn how to eat healthy and stay in shape. Avoid use of tobacco, alcohol or drugs. Do whatever it takes to maximize your health. If you can't walk, run, or swim, someone else may need to risk their life to save yours. If you can, learn how to not be part of the problem, but rather, get the skills needed to help others and be part of the solution.

When stores are closed and city services don't work, you need to know that you have food, water, a first aid kit and other basic supplies. Count the number of people who may be with you during an emergency and figure out what they will need to survive in a three-to five-day period. Buy it, store it, and replace or refresh the supply as needed. (You can find excellent supply checklists on several websites, including *www.TheInnerLink.com.*)

You must also think about what you will need to take with you, in case of evacuation from your home. This includes health records, financial information (such as bank account numbers), insurance policies and school registration information. Make copies of pictures and important videos and DVDs, and document heirlooms, keepsakes and other valuables for insurance purposes. Plan to archive valuables in a secure location, outside of your home. You can also save electronic data to a flash drive, or email data files to yourself, or someone you trust. (If you save the information on a flash drive, you can take the data with you and access it, even if the Internet is not operating. These devices can be attached to a bracelet, necklace, or keychain and transported anywhere.)

List clothing, toiletries, medications and other items that you would want to take with you in the event of evacuation, so that you are able to pack at a moment's notice. Remember to pack relatively light; you do not want to transport heavy bags in the midst of a disaster.

Anticipate the route for evacuation. Plan an additional route that you may follow if your anticipated route is not available.

Response

It all comes down to deciding whether you should stay where you are (which the professionals call "shelter in place,") or evacuate. Devise a plan for each scenario, and store supplies in containers that are appropriate for both in-home and on-the-go use.

You also need to devise a family reunification strategy, in case your loved ones are separated during the course of an emergency. Remember that *where* you go can often be controlled. Select a local meeting place, which you can use in case of a fire or other local, isolated event. Tell family members where this meeting place is, and rehearse your evacuation/unification. Next, select an out-of-region meeting place where you can meet, in person or electronically, if you are separated and your neighborhood is not accessible. Agree to email, cell phone and/or online meeting places in advance.

Recovery

This is a long process and is made much easier if you have planned, prepared and responded successfully. It takes time to recover from an emergency and one often needs help. It is easier to get that help if you have identification, medical information, financial records and insurance information at hand.

If you need emotional, medical or financial support in a time of crisis, do not hesitate to seek it. Lean on your family and friends. Local, state, and federal agencies can make a great difference. Despite the criticism and missed opportunities, the Federal Emergency Management Agency helped hundreds of thousands of people in significant ways. Faith-based organizations can help fill the gap.

Get trained. Remember that when something happens to you, *you* are the first responder; you must respond to the emergency and, whenever possible, prevent it from getting worse. A person can bleed to death while waiting for the EMT to arrive, when simply applying direct pressure to the wound could have saved his or her life. If you wait for the fire truck to arrive before you try to escape a burning building, you risk dying in the fire.

The good news is that training is easy and readily available now. There are many fine training organizations, including the Red Cross, a first-rate organization that has blended online and hands-on instruction with excellent training materials. (To find out where to go for training at the individual, family or organizational level, visit *www.TheInnerLink.com.*)

Think about your community. Just as you have planned for the needs of your family, your school needs to do the same for your children. Be proactive and insist that they have an updated plan that can be shared with first responders. Your place of employment must also plan to protect its workers, and your place of worship must plan to protect its members.

To make your school, workplace and place of worship safer, archive maps and floor plans and share these documents with first responders. (Remember to note important details, such as the location of hazardous or flammable materials.) These institutions need to review their risks, assess their facility and response plans, prioritize their efforts and put resources in place to optimize prevention, preparedness, response and recovery. Planning should include ways that the institution's building can serve the community as a shelter, medication distribution or inoculation center or food service facility, as appropriate.

Community resources need to be identified and shared. The responders need to be able to communicate and cooperate with each other. The community first responder and secondary response (hospitals, National Guard, FEMA) can become overwhelmed, and may need the help of citizens. Many communities rely on volunteer emergency medical technicians and firefighters for their ambulance service and fire departments. Get trained and volunteer. If you have healthcare skills, you should consider creating or joining a Medical Reserves Corps.

Potential threats can be frightening, but you can learn to respond effectively to any emergency. For example, while this book was being written, the authors were consulting with key White House and National Institute of Health researchers to devise a response strategy for an avian flu outbreak. We are pleased to announce that the average citizen can dramatically reduce the spread of disease with simple changes in behavior. (According to a leading NIH researcher, it all boils down to hand washing, covering your mouth when coughing, and avoiding close contact with the public during an outbreak. If the city enforces social isolation or quarantine, having a plan to continue work and school from home, with adequate supplies to do so for a week, will also greatly improve your chances of dealing with H5M1.) The key is informed preparedness. You have taken the first step by reading this book.

We are not claiming that this book will make you an expert at hurricane survival or all-hazards response, but by now, you should be motivated to act on the successes achieved during Hurricane Katrina. Now, you need to help yourself, your family and your community to prepare. As you get started, consider rereading some key preparedness essays, such as those by Kirk Sharp and Nancy Burris. Visit *www.TheInnerLink.com* and click on links to videos and photos provided by this book's editors and writers. Use that site's information to help you plan with your family, and get started with additional training.

Bad things happen, inside and outside of our communities. Some events are in our control, and others are not, but we can take charge of our responses to these events. How can we be sure that a disaster will not emotionally scar our children? How do we recognize the symptoms of post-traumatic stress, in others and ourselves?

In our day jobs (when we are not having a cathartic experience editing this book), Drs. Gillio and Franklin create online and electronic programs that address physical and mental health issues for use by students, teachers, parents, community leaders and concerned citizens. Part of our mission is to educate community professionals and parents to vigilantly watch for emotional health issues in the aftermath of a significant disaster. We wish to make the public aware that these symptoms can manifest years after the event. (As noted earlier, the event does not need to have been personally experienced; trauma can arise simply as a result of repeated exposure to newscasts of disasters.) As our world becomes more uncertain, we may benefit from the following suggestions by Dr. John Tardibuono:

> *As people listen and watch the news, they need to replace thoughts of impending disaster and related feelings of helplessness with thoughts such as: "OK, what can I do to be as ready as possible?" When people make plans it gives them a sense of control...*
>
> *When the event occurs, increased mental resiliency, fueled by realistic, positive self-talk, greatly decreases the likelihood that a traumatic event will be traumatizing to any given individual. The more realistic control individuals have over their feelings and emotions, the more likely they will be able to respond to any event more efficiently. Parents need to be reminded that their emotional reaction to an event plays a major role in how their children are affected.*

You can learn from the essays in this book. Many of the guest authors in this book admitted that it was hard to write these essays, but it was very therapeutic for them. If you, or someone you know, has successfully dealt with this disaster and you wish to share your positive thoughts and images of successes achieved, please contact us through our website (*www.TheInnerLink.com*).

Preparation Summary

- Tear down that silo of authority. Get your government agencies to talk to each other and share plans, rather than compete for precious dollars and resources.

- Train. Use your Red Cross or another source to learn first aid and CPR techniques. Learn how to use an AED.

- Collect and document your basic healthcare information.

- Create a secure list of important non-medical documents and images. You will need them if you evacuate, or if your home is destroyed. Store paper items in a plastic bag or waterproof/fireproof container. Digital information can be stored on a flash drive. These flash drives can be attached to a key ring, for easy access. You can also email electronic information to yourself or someone you trust.

- Create and rehearse a plan for reunification. Select a location in your community, and a location outside your community, where you can meet loved ones in an emergency. Keep a record of email addresses and cell phone numbers of loved ones, and update it as needed. Consider setting up a message board or website where you can communicate with family online.

- Make copies of important personal photos, videos and CD/DVDs. You can either store these copies digitally, using portable electronic devices, or place the copies in a safe location, outside of your home.

- Document your valuable possessions—either by taking pictures, or making lists—and store in a safe place.

- Prepare a list of supplies you will need to survive during an emergency (food, water, batteries, candles, a battery-powered radio, and so on). Buy the supplies, and keep them stocked. (You should have enough to last three to five days, at minimum.)

- Don't wait to plan. Take the first step today.

Response Summary

- Activate your plan

- Stabilize yourself, and then care for others.

- Fix situations that could collapse into a disaster, if unattended.

We thank you for taking the time to read this book. We ask you to celebrate the fact you are alive and think through the successes achieved as you overcame adversities in your own life. Please plan to learn more, and get ready to be a part of the solution in the next disaster or challenge.

Additional practical information and related interviews, photos and stories are available at the Successes Achieved area of www.TheInnerLink.com.

About the Editors

Robert "Dr. Rob" Gillio, MD

Robert Gillio, MD is a Mayo Clinic-trained pulmonologist who left a successful medical practice in the summer of 2000 to found InnerLink, Inc. As that company's chairman, CEO, founder and chief medical officer, he now concentrates his efforts on disease prevention by creating health and safety education programs, such as InnerLink's Coordinated School Health and Safety Solutions including its Wellness series, Projects Breathe, Fitness, Nutrition, Safe, and Recover.

He has held the position of adjunct assistant professor at Penn State University's Milton S. Hershey Medical School. There, he is part of a team that conducts yearly health follow-ups on the 1,800 New York Police Department officers he saw in the immediate aftermath of 9/11. The same team is studying cardiovascular health issues of National Football League players.

Dr. Gillio has coupled his skills as a physician with his quest to make the practice of medicine safer and better understood through technology. A leader in telemedicine, virtual reality surgical simulation, Internet-based education, and case-based learning, Dr. Gillio holds thirteen patents on medical and educational products.

After the attacks in New York City on September 11, 2001, Dr. Gillio treated first responders at Ground Zero and debriefed staff at the Pentagon. In 2002 he published *Lessons Learned from Ground Zero,* an account of his experiences while providing health screenings to NYPD officers. He was subsequently invited to the White House for a series of meetings on related health and safety issues.

His focus on health and safety in schools after 9/11 has led to creation of the My Wellness Projects, a series of online, hands-on programs allowing students to make their own wellness plans. His team has created similar tools for school and community health and safety plans. He is especially proud of his work in supplying the US Surgeon General and faith-based organizations with portal and health record support during Hurricane Katrina relief efforts.

In addition, Dr. Gillio was a founding member of one of the nation's first Medical Reserve Corps, and he built online medical records for use by volunteer doctors who treated the victims of Hurricane Katrina. Creative and innovative, he has started four companies, introducing technology related to reducing drug errors, increasing knowledge delivery, teaching surgical skills, and advancing education. Most recently, he has co-founded the Sudden Cardiac Arrest Foundation, which is focused on empowering citizens to prevent and treat sudden cardiac arrest with tools such as AED's.

A family man, he enjoys spending time with his wife and their five daughters. Since Hurricanes Katrina and Rita, members of the Gillio family have taken twelve trips to the Gulf Coast and New Orleans to do relief and volunteer work.

Evangeline "Dr. Vangy" Franklin, MD, MPH

Dr. Evangeline Franklin, a Henry J. Kaiser Family Scholar, is the director of clinical services and employee health of the City of New Orleans Health Department (NOHD). During Hurricane Katrina, she was co-medical director of the special-needs shelter in the Superdome with NOHD Deputy Director Sandra Robinson MD, MPH. After being rescued, she was evacuated to Dallas with her Superdome coworkers, where she worked with public and mental health organizations in the Dallas/Ft. Worth area, and with the Center for Disease Control in the Convention Center and Reunion Center shelters. In October 2005 she returned to live and work in post-Katrina New Orleans.

In her current position, she is working under the leadership of NOHD Director Kevin Stephens, MD, JD, using emerging technology—such as geographic information systems, or GIS—to reestablish the City of New Orleans' public health and clinical infrastructure. Her areas of responsibility include two NOHD school-based clinics; dental services; and the Women Infants and Children (WIC), Child and Maternal Health, and Healthcare for the Homeless programs. She was the project manager of Health Recovery Week at New Orleans' Audubon Zoo, where portable health records were introduced on a large scale for the first known time. At that outdoor clinic, over five thousand patients received free services and prescription medicines. She continues efforts to further the use of portable health records as project manager of Project Prepare 2006.

She is the author of "Getting People Healthy in New Orleans," a comprehensive monograph that describes the pre-Katrina state of health in New Orleans and the Health Department's agenda for change. She is also the author of a chapter on pre-and post-Katrina healthcare challenges in New Orleans, published in a book titled *There's No Such Thing as a Natural Disaster: Race, Class, and Katrina.*

She has spoken about her experiences at the public health schools of Johns Hopkins and Tulane Universities, as part of the International Society of Hypertension in Blacks annual meeting.

Dr. Franklin has recently spoken at the Texas Association of Public Health and was a panelist at the National meeting of the American Public Health Association meeting.

In addition to her current position at the New Orleans Health Department, Dr. Franklin's career includes: Working as health insurance plan medical director for Aetna and UnitedHealthcare in Louisiana; being named director of urgent care, call center, laboratory, x-ray, physical therapy and occupational medicine services at Cleveland's UniversityMEDNET; an appointment as clinical instructor at Case Western Reserve University School of Medicine; appointments to credentialing and quality committees at University Hospitals of Cleveland and UniversityMEDNET; and a special appointment to the Department of Community Medicine at Tulane School of Medicine. In 1982, she received her medical degree and masters in public health from Yale University and also celebrated her thirtieth reunion as a member of the Class of 1976 at Princeton. She is board certified in internal medicine. Most recently, she has received a Congressional Black Caucus Leadership Award for her heroic work in the Superdome caring for victims of Hurricane Katrina.

978-0-595-41756-8
0-595-41756-6

Printed in the United States
73890LV00004B/303

9 780595 417568